I0060032

# The Rightful Place of Science:
# Frankenstein

# The Rightful Place of Science:
# Frankenstein

Edited by

## Megan Halpern
## Joey Eschrich
## Jathan Sadowski

Consortium for Science, Policy & Outcomes
Tempe, AZ and Washington, DC

THE RIGHTFUL PLACE OF SCIENCE:
Frankenstein

Copyright © 2017
Consortium for Science, Policy & Outcomes
Arizona State University
Printed in Charleston, South Carolina.

The Rightful Place of Science series explores the complex inter-actions among science, technology, politics, and the human condition.

For information on The Rightful Place of Science series, write to: Consortium for Science, Policy & Outcomes
PO Box 875603, Tempe, AZ 85287-5603
Or visit: http://www.cspo.org

Other volumes in this series:

Lloyd, J., Nordhaus, T., Sarewitz, D., and Trembath, A., eds. 2017. *The Rightful Place of Science: Climate Pragmatism*. Tempe, AZ: Consortium for Science, Policy & Outcomes.

Allenby, B. R. 2016. *The Rightful Place of Science: Future Conflict & Emerging Technologies*. Tempe, AZ: Consortium for Science, Policy & Outcomes.

Model citation for this volume:

Halpern, M., Eschrich, J., and Sadowski, J., eds. 2017. *The Rightful Place of Science: Frankenstein*. Tempe, AZ: Consortium for Science, Policy & Outcomes.

ISBN: 0692964177

ISBN-13: 978-0692964170

FIRST EDITION, OCTOBER 2017

# CONTENTS

# INTRODUCTION

Just over 200 years ago, in 1816, five friends gathered on the shores of Switzerland's Lake Geneva for the summer. They were forced to find ways to amuse themselves inside, seeking refuge from the unexpectedly chilly and stormy weather. Ash from the eruption of Mount Tambora in 1815 in the Dutch East Indies (modern-day Indonesia) had created a volcanic winter, what became known as "the Year Without a Summer." Global temperatures fell drastically, causing crop failures and food shortages that led to, among other things, famine, rioting, and looting across Europe and a typhus outbreak in Ireland. The eruption was an act of nature, unforeseeable and ill-understood.

It was against this dramatic global backdrop that the five friends found their summer holiday ruined. Mary Godwin (soon to be Mary Shelley), then just eighteen years old; her married paramour Percy Shelley; the infamous Lord Byron; Byron's personal physician, John William Polidori; and Mary's half-sister, Claire Clairmont, decided to pass the time by writing ghost stories.

Mary Shelley writes that her story, *Frankenstein; or The Modern Prometheus*, came to her in a flash of inspiration after listening to a conversation between Byron and Percy Shelley about the origins of life. Their discussion of the science of the day provided the spark for one of the most enduring horror stories in western culture. Shelley's tale follows Victor Frankenstein, a scientist driven by his obsession to create life. Frankenstein succeeds in creating life but, when faced with his creation, he runs, abandoning the creature to a world that cannot accept him. The creature seeks love and companionship, but his frightful appearance strikes fear in the hearts of everyone who sees him. In the end, this rejection not only from society at large, but also from Victor, fills the creature with rage. When Victor denies him a companion, he destroys everything his creator loves. As the literary world marks the 200th anniversary of the story's writing and publication, we, like many before us, want take some time to reflect on what Mary Shelley's tale can teach us.

In 2012, the legendary sociologist and philosopher of science and technology Bruno Latour, who is widely known for his studies of scientific practice, took up the Frankenstein myth as a way of exploring sustainability. Latour implored us to "love our monsters" and to treat them as we would our own children. In his essay, he provides an interpretation of the novel that requires the reader to consider not the act of creation, but the abandonment of that creation as Victor Frankenstein's most egregious offense. Shelley's story, Latour tells us, is a cautionary tale for our age:

> *Let Dr. Frankenstein's sin serve as a parable ... [a]t a time when science, technology, and demography make clear that we can never separate ourselves from the nonhuman world — that we, our technologies, and nature can no more*

*be disentangled than we can remember the distinction between Dr. Frankenstein and his monster ...*[1]

This observation from Latour about our connectedness to our technologies and to nature indicates why *Frankenstein* is still such a compelling story, and why the novel merits its own volume in *The Rightful Place of Science* series. Latour's challenge to love our monsters is echoed in David Guston's call for responsible innovation, which is explored in his chapter, "The Insufficiency of Cool" and in much of the work done by the School for the Future of Innovation in Society, which Guston directs. As we humans continue to wrestle with what we create through innovation and through neglect, *Frankenstein* offers rich ground for exploring the literary, scientific, social, and cultural dimensions of creativity and responsibility.

This volume's explorations of *Frankenstein* provide a series of lenses through which the authors take up the theme of creativity and responsibility, in the novel and in the world, in very different ways. The chapters range from close readings of Shelley's 1818 publication to explorations of larger Frankensteinian themes. Some authors draw out new interpretations from the structure and story, while others focus on Victor Frankenstein's character, offering different interpretations of his mistakes as a human and as a scientist. Still others turn their attention to the countless adaptations, allusions, and cultural reverberations of the story. Finally, some of the authors provide insight into creations emerging in our current moment in history: synthetic biology, smart cities, and the cultural text as a network.

---

[1] Bruno Latour, "Love Your Monsters: Why We Must Care for Our Technologies As We Do Our Children," *The Breakthrough Journal* (Winter 2012).

In Chapter 1, Joey Eschrich examines the novel's literary roots and style. He begins by considering Shelley's references to John Milton's *Paradise Lost*. He draws out the parallel between Satan's rage at God in Milton's epic poem and the creature's rage at Victor Frankenstein. Eschrich then explores Mary Shelley's intellectual and literary influences, including her parents, Mary Wollstonecraft and William Godwin, and her literary companions that summer. Finally, he looks at the structures of the novel, and especially at the use of written correspondence between characters, as a way of unfolding the narrative and revealing relationship dynamics.

In Chapter 2, Bina Venkataraman examines the overused metaphor of "playing God" in the context of Shelley's novel and in culture. She argues that the admonition against "playing God" hides the more complex responsibilities of scientists and innovators to ensure that their creations are integrated into society. Innovators, Venkataraman asserts, must participate in robust public debate about the possible consequences of scientific and technological transformation.

Megan Halpern draws on feminist theory in Chapter 3 to consider the different ways readers understand Victor Frankenstein's creation, and how he understands himself. She describes three frames for interpretation: the creature as a technology, as a monster, and as a human. How readers choose to interpret the creature shapes the meanings they take from the text. Each of these interpretations has its own implications, but choosing to understand Frankenstein's creature as a human provides insight into how it feels to be ostracized by one's own community, and ultimately what it means to be marginalized by society.

Charlotte Gordon, in Chapter 4, suggests that Shelley's story rests on Frankenstein's three big mistakes: isolation, neglect, and lack of funding. Modern artificial intelligence researchers and other innovators would be well advised to learn from these mistakes, offering hope for future Frankensteins.

In Chapter 5, Sara Walker, a physicist and astrobiologist, asks us to see Victor Frankenstein as a scientist. Walker suggests that the questions that drove Frankenstein to build his creature are those that motivate all scientific research. This chapter provides a history of research into spontaneous generation, noting that scientists still have many questions about what life actually is — let alone any answers as to how to create it. But the difference between the scientists Walker writes about and Victor Frankenstein isn't that they failed where he succeeded. It is that they persisted in their investigations where he gave up and stopped asking questions. She proposes that the creature's descent into violence and murder are an indication of Victor's failure as a scientist: this is what happens when scientists lose their curiosity and their capacity for research.

Continuing the theme of *Frankenstein* as an exploration of science, Keven Esvelt, David Guston, and Alyssa Sims focus on specific scientific advances and their potential for use and abuse. Esvelt's chapter takes on the popularity of CRISPR technology as a genetic tool, while Guston focuses on synthetic biology. Esvelt suggests that both reality and fiction suggest that scientists need to open up their work to public scrutiny and input. Guston discusses the motivations and justifications for pursuing cutting-edge scientific advancements in synthetic biology. He asserts that coolness is an insufficient rationale for taking on a creative scientific project and argues that socially responsible research and innovation must include foresight. In Chapter 8, Sims suggests we see Frankenstein not simply as a scientist, but as a pioneer of DIYbio and biohacking as well.

The final chapters consider the broader cultural implications and legacy of *Frankenstein*. In Chapter 9, Jathan Sadowski considers the contemporary "smart city" movement through the lens of *Frankenstein*. He looks at the ways the technologies and systems that comprise the smart city can shape these cities, as well as the people who inhabit them. Like Frankenstein's creature, he writes, the technologies in these cities have the potential to master their human inhabitants and creators. But unlike in Shelley's novel, the creators may not be the ones who pay the price for decisions made without concern for unintended consequences.

Finally, in Chapter 10, Ed Finn looks at *Frankenstein* as a network text. The original novel, he tells us, has become part of a tangled web of content across an incredible variety of genres and media forms. That network of stories, images, and representations, much like the creature himself, has taken on a life of its own. This network provides opportunities for a struggle over meaning, but also for playfulness and collaborative creation. Finn suggests that we need this network to keep the myth of *Frankenstein* alive, and in so doing, to keep the questions *Frankenstein* raises about creativity and responsibility at the forefront of our collective consciousness.

This volume's collection of essays was written to join the chorus of bicentennial thinking about the novel. The authors in these pages examine the roots and origins of Mary Shelley's tragically flawed scientist and his benighted monster. They consider Shelley's tale as a parable of creativity and responsibility that might help us better understand our current creative dilemmas. And they point toward the ways that Shelley's text foreshadows future technological innovations, and the dilemmas we anticipate from emerging fields such as synthetic biology and artificial intelligence. The authors' hope is that 200 years after its inception, this story of a scientist who failed to care for his masterpiece can help us learn to care for our own.

When we read *Frankenstein* through the lens of creativity and responsibility, we can't help but see Victor's failure to take responsibility for his creation as his downfall. But that isn't enough to help us understand our current and future challenges. To do that, we must push ourselves further. It has been easy to see Victor's hubris as a reason not to innovate, to use the precautionary principle to avoid risks, even when those risks might prove beneficial. But this is not the lesson Bruno Latour hopes we take from the tale, nor is it the lesson you will find in these pages. Our challenge is to continue to value the process of innovation as one of humanity's greatest characteristics, while developing the attention, commitment, and care to our innovations that Victor lacked.

*— Megan Halpern, Joey Eschrich, and Jathan Sadowski*
*October 2017*

# 1

## UNSETTLING *FRANKENSTEIN*

## Joey Eschrich

We all know the iconic representation of Victor Frankenstein and his creature. A mad scientist, his white lab coat askew as he leans over his progeny. An inert body atop a metal table, swathed in a sheet. Strikes of lightning and claps of thunder, loud and bright and immediate. Incomprehensible machinery towering over the table, perhaps lining the walls of the room. Sparks and arcs of electricity dance and crackle. Maybe the doctor is wild-eyed, his nerves frayed, his eyes sunken. Perchance he cackles.

These images, whether we encounter them on cereal boxes, in Saturday morning cartoons, or in serious art films, tell us the same story about *Frankenstein*: it's a fable about hubris, a warning about the dangers of out-of-control science and innovation. The creature is an ungodly creation—that is, a human creation—so the results of the experiment are predictably calamitous.

Perhaps the exemplar of *Frankenstein*-as-hubris-parable is James Whale's 1931 monster movie, *Frankenstein*. The film, especially its legendary "It's alive!" scene, provides a crystal clear example of how Mary Shelley's story has been adapted to encourage us to think of Victor's tragic mistake as hubristic scientific overreach.

If you watch a clip of the "It's alive!" scene you'll find that it feels familiar, even if you've never seen this particular movie. The first thing you'll notice is Colin Clive, the actor who plays Victor. Clive gives an unforgettably unhinged performance, capped off with a perfectly delivered line, vibrating with hysteria and grandiosity: "Now I know what if feels like to be God!" Whale, the director, instructed Clive to portray Victor as "an intensely sane person" who is nonetheless "at times rather fanatical," his "nerves … all to pieces."[1] This Victor Frankenstein has passed from Enlightenment rationality into the pre-rational, atavistic realm of alchemy and incantation, snatching the power of creation from the hands of God.

The scene features dramatic shifts between long shots in which we see the intimidating electrical machinery of creation—complete with giant orbs and serpentine coils—and close-ups of grimacing faces and the twitching fingers of the creature emerging into life. None of this gadgetry is depicted in the novel; Shelley deliberately leaves the process of creation unseen. In Whale's film the creature's awakening takes place in a cavernous, machine-filled womb. The insistent roar of thunder threatens to drown out the dialogue. And the pacing of the scene is deliciously ritualistic: we see the creature's metal bed slowly raised, then lowered, in two patient, unbroken camera shots.

All of these decisions about camerawork, pacing, acting style, dialogue, and set dressing clearly point toward a reading of the scene as a moment of catastrophic hubris. This interpretation of *Frankenstein* as a warning against adventurous innovation and self-aggrandizing creativity

---

[1] Susan Tyler Hitchcock, *Frankenstein: A Cultural History* (New York: W. W. Norton & Co., 2007), 147.

has become an ingrained part of our cultural mythos. But in her original 1818 text, Shelley marshals style in a very different way to create a more nuanced fable. Careful attention to Shelley's stylistic choices reveals that the central theme is a more subtle exploration of the connections between scientific creativity and societal responsibility.

Reframing *Frankenstein* as a story about creativity and responsibility, rather than scientific hubris, is a particularly helpful way to account for its strikingly odd style and structure. *Frankenstein* is weirder than you might imagine—especially if what you're expecting is a straightforward Gothic horror story about the mad scientist, his electrified laboratory, and his green, hulking, bolt-necked creation.[2] Shelley unfolds the novel's narrative in nested layers, leads the reader on lengthy narrative tangents that seem to do little to advance the plot, presents the full text of letters exchanged between characters, and deploys dense thicket of allusions and references to other texts both literary and philosophical. Her stylistic decisions make the novel an interpretive puzzle, not the pat, didactic story that pervades popular culture.

*Frankenstein*'s weirdness only makes sense if we view it as a way of demonstrating, through style as much as content, the responsibility of creators to care for the products of their creativity. In Shelley's story, Victor fails not as an innovator, but as a caregiver and parent. He fails to situate his scientific work in a social context, and thereby denies the creature the resources he needs to remain sane and healthy, to thrive and be a productive member of a community.

Shelley's use of stylistic weirdness to unsettle a simple interpretation and plunge us into murkier ethical waters

---

[2] Christopher P. Toumey, "The Moral Character of Mad Scientists: A Cultural Critique of Science," *Science, Technology, & Human Values* 17, no. 4 (1992): 411-437.

resonates with the literary scholarship on "the weird." Literary weirdness involves using fantastical tropes (for instance, regenerating dead flesh into a living, reasoning creature) in a way that unsettles and challenges the reader, instead of using them to construct a "consolatory" story that comfortably reaffirms the reader's values and worldview.[3] In this sense, weirdness helps to explain *Frankenstein*'s unique staying power over two busy centuries of adaptation and critical discourse:[4] it troubles our attempts to reduce its themes, and the turmoil of its characters, to a simple, reassuring ethical calculus. It implicates us as creators and users of technology, challenging us to question our assumptions about what it means to be a responsible person in a thoroughly technological world.

But wait a second. Why is it important to go back to the novel and interpret it now, two centuries after its original publication? Here's the thing: we're at a moment in scientific and technological history where our capabilities in obviously Frankensteinian areas like genetics, artificial intelligence, robotics, and tissue engineering are exceeding our ability to make sense of them, to sensibly regulate them, and to judge their risks. Critics of technological change have long interpreted and portrayed *Frankenstein* as a cautionary tale, a dire warning to stop "playing God" before it's too late. But with formidable economic and political forces driving technological advancement, scientific and technical knowledge will continue to progress.

So myths about hubris aren't what we need. They can't work in the face of breakneck innovation, and they don't

---

[3] Darja Malcolm-Clark, "The New Weird," in *Women in Science Fiction and Fantasy, Vol. II*, Robin Anne Reid, ed. (Westport, CT: Greenwood Press, 2009), 234-236.

[4] *Frankenstein* is the most-assigned novel on U.S. college syllabi, according to Columbia University's Open Syllabus Project (http://explorer.opensyllabusproject.org).

give us the tools to understand how to deal with scientific and technological change when it happens. Rather than the tale many of us think we know, Shelley's actual story is a potent myth about our duties as creators, as caretakers who must not neglect our creations. It's just the kind of story about invention and its consequences that could prompt more ethical behavior on the part of innovators and policymakers, and remind the rest of us just how important our attention and participation really is.

🔲   🔲   🔲

Shelley's first key stylistic choice is to situate her novel in relationship with John Milton's epic seventeenth-century poem *Paradise Lost*. Milton retells the Biblical story of Adam and Eve, the Garden of Eden, and the rebellious angel Satan—with a shocking amount of sensitivity to Satan's perspective. *Paradise Lost* positively haunts the 1818 edition of *Frankenstein*: Shelley uses a quote from the poem as the book's epigraph,[5] and the text is studded with allusions to it.

Most importantly, *Paradise Lost* is one of the books that the creature serendipitously finds in the forest after his abandonment by Victor. The creature uses the book to educate himself about human relations, history, religion, and morality.[6] *Paradise Lost* gives the creature a moral prism through which to understand himself as wronged and abused by his creator—as the protagonist in his own story. The relationship between Milton's anguished, articulate, and eventually vengeful antihero Satan and his emotionally distant creator, God, serves as a template for the dynamics between the creature and Victor.

---

[5] "Did I request thee, Maker, from my clay / To mould me man? Did I solicit thee / From darkness to promote me?"

[6] The other two books the creature finds are *Plutarch's Lives* and Goethe's *The Sorrows of Young Werther*.

In *Paradise Lost*, the conflict between God and Satan stems from God's unwillingness to meet the emotional, intellectual, and spiritual needs of his angels. He creates these beings with unsurpassed intelligence and sensitivity, then relegates them to an eternity of worshipful servitude in heaven. At the root of Satan's rage is God's unwillingness to recognize him and the other angels as individuals with needs and thoughts of their own, not merely messengers and gleaming symbols of divine glory. Milton's God fails to create the conditions for Satan and the other angels to thrive, to use their endowments constructively, to reach their full potential.

Victor is similarly unable to see the creature as a being with integrity and an inner life of his own. He is comfortable with the creature as abstract proof of his scientific mastery and godlike powers of creation, but as soon as the creature actually begins to *be*, Victor exits the scene in disgust and horror. He is only ever a hideous abomination in Victor's eyes, even when he eloquently pours out his heart to Victor and begins to deliver ultimatums. Once the creature's rampage of violent retribution has begun, Victor continues to be emotionally callous: going back on his promises, running and hiding, and stalling without explanation. Victor never acts as if he's dealing with a fellow human.

Like Milton's God, Victor fails to situate the creature in a social context in which he can thrive. He denies the creature access to all of the social institutions and bonds that give life meaning and provide support when we are in need: family, friendship, education, civil society, and religion. Victor's decision to conduct his experiments in secret, hidden from his family and colleagues, geographically dislocated from his home and his university, makes it impossible for his creature to gain access to the resources he needs to construct a livable life.[7]

The ideas of Shelley's famous parents, the philosophers Mary Wollstonecraft and William Godwin, are another rich source of allusions and references throughout the text of *Frankenstein*. Mary Shelley (Godwin, when she first conceived of the *Frankenstein* story) and her intellectual circle—including her soon-to-be husband Percy Shelley and the temperamental and outrageous aristocrat, poet, and radical Lord Byron—shared a reverence for Wollstonecraft, whose *A Vindication of the Rights of Woman* (1792) is a foundational text for Western feminist thought.[8] In *Vindication* and other writings, Wollstonecraft argued for the importance of education and the structure of the family. These institutions could shape people into rational adults and good citizens able to take care of themselves, comport themselves with virtue, and contribute to the overall well-being of society.

William Godwin, in his most influential book, *Enquiry Concerning Political Justice and its Influence on Morals and Happiness* (1793), argues that "character is a function of experience and that the type of government under which people live has an overwhelming impact on their experience—bad government produces wretched men and women."[9] Both of Shelley's parents were committed to "social constructionism": the idea that social forces, not an interior essence, molds us into who we are and forms a

---

[7] Judith Butler, *Frames of War: When Is Life Grievable?* (New York, NY: Verso, 2009), 20-23.

[8] Charlotte Gordon, *Romantic Outlaws: The Extraordinary Lives of Mary Wollstonecraft and Her Daughter Mary Shelley* (New York, NY: Random House, 2015), 131, 159, 286-7.

[9] Mark Philp, "William Godwin," *The Stanford Encyclopedia of Philosophy*, Edward N. Zalta, ed. (Summer 2013 Edition): http://plato.stanford.edu/archives/sum2013/entries/godwin.

structure for our personality, morality, social relations, and belief systems. In other words, Frankenstein's creature isn't born wrathful and violent; he's made that way by his abandonment, isolation, and the rejections he suffers.

Wollstonecraft's and Godwin's ideas provide the philosophical threads connecting the creature's trials and tribulations with his character development. He needs socialization to thrive, but he is abandoned by his creator. He despairs, and almost dies alone. He finds great works of literature and enriches himself through learning. He spies on a loving family and learns about emotional support and the dynamics of loving families through observation. He is rejected by the people he tries to befriend and form bonds with. These traumatic rejections cause him to become embittered and turn against his neglectful creator. Tragedy ensues for all.

Wollstonecraft famously inveighed against a British society that she perceived as obsessed with appearances, which stunted the moral and intellectual development of people, especially women. This is a minor but important connection to *Frankenstein*: while in contemporary adaptations we're used to seeing an inarticulate or mute creature, in Shelley's novel the creature speaks beautifully, in heartfelt soliloquies, about the pain of being rejected, sometimes violently, merely because of his grotesque physical appearance. Just as Milton did for Satan in *Paradise Lost*, Shelley chose to make the creature the most insightful and poetic speaker in the story.

Shelley did not adore her father with the same unbridled intensity as she did her late mother, who died shortly after giving birth to her. Shelley and Godwin were estranged for years before and after the publication of *Frankenstein*. Her dedication of the novel to Godwin was an attempt to mend fences after years of vitriol and hurt feelings stemming from her decision to run off with Percy

Shelley, a married man with a child. It's hard not to see Mary reflected in her wounded creature, desperate for love, affection, and guidance from a brilliant but disdainful father.

*Frankenstein* also marks a key point of intellectual disagreement with Godwin's philosophy. While Godwin believed in human perfectibility—the idea that individuals would always become more enlightened and moral as their knowledge and understanding increased—*Frankenstein* strikes a note of skepticism. Victor unlocks the secrets of life, death, and regeneration, but he doesn't develop morally as his knowledge expands. When the creature's wrath threatens his loved ones, Victor fails to come clean and share his terrible secret. His hidden shame kills the people he loves, and for most of the novel he takes no decisive steps to put an end to it.

Taking creativity and responsibility as *Frankenstein*'s central theme helps to give meaning to the novel's mysterious nesting doll structure: Victor's story is embedded in a narrative about a voyage to the North Pole, and the creature's account of his early life is embedded in Victor's narrative. The novel begins with Robert Walton, the captain of a scientific expedition to the North Pole. Walton is an earnest young adventurer and a clear parallel to Victor, who also sought fame and prestige through a risky scientific endeavor. Stuck en route to the Pole, waiting for the ice to thaw, Walton and his crew pick up a haggard and raving Victor, who recounts his tragic life story.

Walton's choice about whether to press on towards the North Pole in the face of danger and possible mutiny, or to turn back and allow his crew to return home unharmed, becomes its own investigation into creativity and responsibility. Are the untold wonders of the Pole, and the prestige of reaching it, worth the existential risk? Victor gives Walton highly contradictory and unhelpful advice: in one breath, Victor warns against the dangers of excessive ambition, and in the next exhorts Walton to seek glory at any cost. It's clear that he hasn't learned much from his multi-year campaign against his creature, who he still calls a "demon" instead of recognizing him as a thinking, feeling, and neglected being. Walton's story clarifies the moral stakes of the interplay between scientific inquiry and social responsibility. When Walton comes through for his crew in the end, and (as far as we know) returns home safely, he provides a foil for Victor's disastrous decisions.

What's particularly important is how Walton reaches his decision to abandon his quest for the Pole: he talks to people. He shares his thinking about ambition and risk with Victor, he trades letters back and forth with his sister, and he talks to his crew. Even when things are momentarily dicey and the crew threatens to mutiny, Walton keeps lines of communication open, and he eventually makes the cautious and responsible decision. Victor fails completely at sharing with others his scientific ambitions and his thinking about risk and responsibility. He isolates himself from his colleagues and family to conduct his experiments; even when people begin to die, he stubbornly keeps his secret. The conduct, not the content, of Victor's work dooms him. He's not excessively ambitious but excessively secretive, which makes him irresponsible. It's his failure to embed his scientific endeavors in a social context that leads to his downfall.

*Frankenstein*'s many tangents, which don't seem to move the narrative forward, are another aspect of the novel's weirdness. These tangents can be jarring for readers who are familiar with the story as it's told in most adaptations. Shelley's stylistic choice to meander repeatedly from the main thrust of the plot makes more sense, and becomes more interesting, if we see these tangents as efforts to add depth to the theme of creativity and responsibility.

One of these tangents centers on young Justine Moritz, who is taken in and treated exceptionally well by the Frankenstein family as a servant after her mother's death. We learn quite a bit about Justine and especially about her life history before she joins the Frankenstein household, which is surprising, given her minor role in the plot. But Justine's story is important because it acts as a counterpoint to Victor's treatment of his creature: she is embraced by her adoptive family and valued as a human being, while the creature is scorned by everyone he encounters. Justine is young and beautiful, and that's the key difference: no one, especially Victor, can look past the creature's hideous appearance.

*Frankenstein*'s social constructionist thinking—inherited from Wollstonecraft and Godwin—is in evidence here too. The desperate, despondent creature lashes out violently against his utter rejection. He kills young William Frankenstein, Victor's brother, after the young boy panics at the sight of him, and he then frames Justine for the murder. Sweet Justine, loved by the Frankenstein family, selflessly admits to a crime she did not commit, in a frenzy of Catholic guilt and a mistaken belief that she is cursed, leading to her execution as punishment for William's murder. The creature's dark social history makes him into a murderous fiend, while Justine's exposure to religion at a young age inclines her toward crippling, self-destructive guilt.[10] Both the creature and Justine are products of their environment; attending to Justine's life in detail helps emphasize the way that the creature's life shapes his behavior and destiny.

---

[10] Mary Shelley and her husband Percy were noted atheists; Percy's first large-scale work of poetry, *Queen Mab* (1813), was incendiary because of its antagonistic stance on religion.

A second illuminating tangent is Victor's near-disaster in Ireland. He is despised there by villagers who are suspicious that he murdered his best friend, Clerval, who in fact was strangled by the creature. In this case, Victor is cared for by the local magistrate Mr. Kirwin, who treats him with sympathy and compassion. Kirwin helps to arrange Victor's defense, converses kindly with him, and arranges for him to recuperate from his mental and physical breakdown in the best room of the village jail. Like the Justine story, Victor's encounters with Kirwin do little to advance the plot. But this episode underlines the importance of human kindness and acceptance for people's well-being and sanity. The modest gestures of kindness the distraught Victor receives from Kirwin are instrumental in keeping him from losing his mind. It's clear, by extension, that the creature's despair and fury could have been tempered by just one small glimmer of understanding and care.

The Justine and Kirwin episodes demonstrate the importance of structured and institutionalized human relationships — the family, a humane justice system — for nurturing human life. Victor brought his creature to life with no plan for how he would be integrated into society. In fact, the way Victor conducted his experiments — hidden from his friends, family, and fellow scientists — dooms the creature to isolation. In order to survive, we need to be socially legible: to have a place in the world, a visible identity, membership in social groups, and connections with institutions to support us, safeguard our rights, and care for us when we are in need. Justine and Victor had these social connections, whereas the creature is forever alienated from society because Victor created him to without also creating a structure in which he could thrive.

The theme of creativity and responsibility also gives new meaning to *Frankenstein's* epistolary structure. Long, tender letters exchanged between loved ones continuously interrupt the narrative's momentum. The very beginning of the novel, in fact, is a letter from the polar explorer Walton to his sister Margaret, and *Frankenstein* ends with a series of missives from Walton to Margaret. The letters are full of personal details, endearments, and details that do nothing to advance the plot. This just can't, you think, be the most efficient and clear way to tell this story. But that's exactly it—this structure isn't chosen for efficiency and clarity. Instead, the letters are tangible artifacts of the social ties, intellectual exchanges, and mutual obligations that are the building blocks of lasting, fulfilling relationships with family and friends. The letters offer a glimpse into the ways that the novel's characters, excepting the creature, are ensconced in rich social worlds and networks of support and consideration.

The letters demonstrate the investment of time, wit, and emotional energy that makes human relationships functional and valuable. The long description of how the creature acquires language, through abandoned books and eavesdropping, underscores just how far he is from any form of nurturing social interaction. Instead of learning language from loved ones, he is forced to scavenge it. The other characters are sustained emotionally, often in the midst of immeasurable trauma, by the support they receive from one another.

Meanwhile, the creature is rarely able to elicit even a neutral, nonviolent reaction from another person. The letters reveal precisely what the creature is missing. They embody the ties of mutual responsibility that make life bearable, that keep us all from sinking into depression, destructiveness, or obsession. It's critical that Walton, who narrowly avoids being the victim of his own delusions of grandeur, is in continuous contact with his sister, who lovingly discourages him from betting everything on his quest for the North Pole. Perhaps if Victor had shared more of his ideas, passions, and plans with his loved ones, he too could have averted disaster and wild-eyed death on the desolate Arctic ice.

*Frankenstein*'s weirdness is no accident. Shelley's stylistic choices place us in an ethical landscape where being a creator is complicated. The consequences of our decisions radiate out unpredictably and affect others. Our responsibilities extend far beyond the laboratory and the moment of creation. And we are all, if we hope to avoid Frankenstein's fate, tangled up in relationships that give our actions meaning.

# 2

## THE PROBLEM WITH "PLAYING GOD"

## Bina Venkataraman

In Mary Shelley's *Frankenstein*, the notorious creature is hiding from human view when he encounters a suitcase in the woods filled with books and clothing. The creature reads Milton's *Paradise Lost* and can't help but compare himself to both Adam and a fallen angel. He recounts his discovery to his maker, the distraught Victor Frankenstein, with indignation:

> *Accursed creator! Why did you form a monster so hideous that even YOU turned from me in disgust? God, in pity, made man beautiful and alluring, after his own image; but my form is a filthy type of yours, more horrid even from the very resemblance.*[1]

Since its publication nearly 200 years ago, Shelley's gothic novel has been read as a cautionary tale of the dangers of creation and experimentation. James Whale's 1931

---

[1] Mary Shelley, *Frankenstein; or The Modern Prometheus*, ed. D. L. Macdonald & Kathleen Scherf, 3rd Ed. (Peterborough, Ontario: Broadview Press, (2012) [1818]), 144.

film took the message further, assigning explicitly the hubris of playing God to the mad scientist. As his monster comes to life, Victor, played by Colin Clive, triumphantly exclaims: "Now I know what it feels like to be God!"

The admonition against playing God has since been ceaselessly invoked as a rhetorical bogeyman. Secular and religious, critic and journalist alike have summoned the term to deride and outright dismiss entire areas of research and technology, including stem cells, genetically modified crops, recombinant DNA, geoengineering, and gene editing. As we near the two-century commemoration of Shelley's captivating story, we would be wise to shed this shorthand lesson—and to put this part of the *Frankenstein* legacy to rest in its proverbial grave.

The trouble with the term arises first from its murkiness. What exactly does it mean to play God, and why should we find it objectionable on its face? All but zealots would likely agree that it's fine to create new forms of life through selective breeding and grafting of fruit trees, or to use in-vitro fertilization to conceive life outside the womb to aid infertile couples. No one objects when people intervene in what some deem "acts of God," such as earthquakes, to rescue victims and provide relief. People get fully behind treating patients dying of cancer with "unnatural" solutions like chemotherapy. Most people even find it morally justified for humans to mete out decisions as to who lives or dies in the form of organ transplant lists that prize certain people's survival over others.

So what is it—if not the imitation of a deity or the creation of life—that inspires people to invoke the idea of "playing God" to warn against, or even stop, particular technologies? A presidential commission charged in the early 1980s with studying the ethics of genetic engineering of humans, in the wake of the recombinant DNA revolution, sheds some light on underlying motivations. The commission sought to understand the concerns expressed by leaders of three major religious groups in the United States—representing Protestants, Jews, and Catholics—who had used the phrase "playing God" in a 1980 letter to President Jimmy Carter urging government oversight. Scholars from the three faiths, the commission concluded, did not see a theological reason to prohibit genetic engineering.[2] Religious scholars' concerns, it turned out, weren't exactly moral objections to scientists acting as God. Instead, they echoed those of the secular public; namely, they feared possible negative effects from creating new human traits or new species. In other words, the religious leaders who called recombinant DNA tools "playing God" wanted precautions taken against bad consequences but did not inherently oppose the use of the technology as an act of human hubris.

What seems to drive most contemporary critics who invoke "playing God" is, likewise, not a religious or moral objection to human beings playing the role of creators, but a fear of the unintended social consequences of scientific discoveries and new technologies. (Such fear finds its footing in historic examples of chemicals deployed as weapons or leaked into drinking water, life-saving drugs that benefitted only the wealthy, and unethical experiments such as the Tuskegee Syphilis Study.) To urge against playing God, moreover, is to convey a mistrust of scientists—and to criticize their arrogance in the face of the power and unpredictability of nature. The phrase has become a stand-in for these deeper sources of public discom-

---

[2] President's Commission for the Study of Ethical Problems in Medicine and Biomedical and Behavioral Research, *Splicing Life: A Report on the Social and Ethical Issues of Genetic Engineering with Human Beings* (Washington, DC: US Government Printing Office, 1982).

fort with science and technology that are better exposed and examined, rather than cloaked in superstitious warning.

The late evolutionary biologist Stephen Jay Gould once argued that Hollywood had "dumbed down" the subtleties of the original *Frankenstein*.[3] Whale's film, he noted, reduces the creature's murders to biological determinism: it's because his creator gave him the brain of a former criminal. While the popular movie attributes the creature's violence purely to nature, the novel makes clear that it comes from the rejection he experiences from Victor Frankenstein and the rest of humanity. The wisest warning that Shelley proffers is not against creating life or imitating God, but rather against neglecting the outcomes of experimentation and discovery. Victor, abhorred by the hideousness of his creature, cruelly abandons his invention, leaving him without the care and education to become a moral being. His murderous rampage is the result not of having been invented in the first place but of profound neglect.

The lesson for contemporary science, then, is not that we should cease creating and discovering at the boundaries of current human knowledge. It's that scientists and technologists ought to steward their inventions into society, and to more rigorously participate in public debate about their work's social and ethical consequences. *Frankenstein*'s proper legacy today would be to encourage researchers to address the unsavory implications of their technologies, whether it's the cognitive and social effects of ubiquitous smartphone use or the long-term consequences of genetically engineered organisms on ecosystems and biodiversity.

---

[3] Stephen Jay Gould, "The monster's human nature," *Natural History* 103, no. 7 (1994): 14.

Some will undoubtedly argue that this places an undue burden on innovators. Here, again, Shelley's novel offers a lesson. Scientists who cloister themselves as Victor did — those who do not fully contemplate the consequences of their work — risk later encounters with the horror of their own inventions. (Albert Einstein, who contributed only indirectly to the making of the atomic bomb, tried to avoid this fate in a famous letter to President Franklin Roosevelt,[4] while J. Robert Oppenheimer grew regretful after making the bomb.) Scientists who do not engage in public debates about their research may face backlash that curtails the technologies themselves, as we've seen in European bans on genetically modified organisms. Conscientious scientists will take on such social risks as engineering challenges, building safer self-driving cars and algorithms that correct for, rather than replicate or exacerbate, human bias and discrimination.

The environmentalist and futurist Stewart Brand opened the first *Whole Earth Catalog* in 1968 with this line: "We are as gods and we might as well get good at it." The statement was a reflection on humanity's awe-inspiring power to change the planet and the tragedy of the environmental impact it had already wrought. (Brand later wrote that he "stole" the line from the related words of the British anthropologist Edmund Leach.) The mantra "we might as well get good at it" could serve to expand the metaphor and lessons of *Frankenstein* for our time, offering a ready response the next time "playing God" surfaces in popular dialogue. And whether it's artificial intelligence, gene editing, or some other new technology on the horizon, that should happen any minute now.

---

[4] Walter Isaacson, "Chain Reaction: From Einstein to the Atomic Bomb," *Discover* (March 2008).

# 3

## THE CREATURE IS MORE HUMAN THAN THE CREATOR

## Megan Halpern

*"Unfeeling, heartless creator! You had endowed me with perceptions and passions, and then cast me abroad an object for the scorn and horror of mankind."*[1]

These words, spoken by Victor Frankenstein's creation, are part of a monologue that spans multiple chapters in the novel. The creature's desperation and devastation at his own condition ring through his appeals and threats to Victor to create a female companion for him. He continues: "But on you only had I any claim for pity and redress, and from you I determined to seek that justice which I vainly attempted to gain from any other being that wore the human form."[2] This monologue details the pain and hope of the creature, and though he is the stuff of monster movies, in this moment he is articulate and deeply emotional.

---

[1] Mary Shelley, *Frankenstein; or The Modern Prometheus*, ed. D. L. Macdonald & Kathleen Scherf, 3rd Ed. (Peterborough, Ontario: Broadview Press, (2012) [1818]), 152.

[2] Shelley, 152.

Victor's initial flight from his creation, and his subsequent abdication of responsibility for the creature's actions, signal his failure as a scientist and as a father. Meanwhile, the creature, whom Victor deems a monster and demon, displays a greater capacity for understanding humanity than do the humans he encounters throughout the tale.

There are a number of ways we can choose to see Victor Frankenstein's creation. Modern incarnations of the Frankenstein myth often interpret the creature as a dangerous invention. Like popular perceptions of genetically modified foods or nanotechnologies that have unintended consequences, the creature is a thing: an artifact that wreaks havoc and leaves destruction in its wake. The humans in Mary Shelley's original story see a grotesque monster; they understand him as a demon who commits unspeakable acts of violence and murder. Finally, though he is not afforded the luxury of being treated as such by anyone he encounters, he sees himself as human.

What happens when we also view the creature as human? How does this change our understanding of Shelley's text, and the lessons we can learn from Victor's choices? What happens when we identify the creature as the protagonist in a tragic tale of isolation and despair? Though the creature begins his life without malice, first his hideous appearance, and eventually his own antisocial actions separate him from society, leaving him to seek opportunities to connect in whatever ways he finds available to him. These grow from hiding in the shadows to blackmail and threats, and ultimately, to violence and murder. These characteristics and actions may be read as those of a non-human technological artifact or of a monster. But when we see them as human actions, we are left with critical questions about how social interactions shape individuals, how we distribute accountability for social actions, and who bears responsibility for the actions and reactions of those who have been marginalized.

## Creature as Technology

Many modern incarnations of *Frankenstein* and Franken-metaphors leave aside the creature's consciousness and agency, using the name for any kind of new technology with the potential for unintended consequences. Genetically modified foods have earned the title "Frankenfoods" in debates over their safety; Frankenstein is a pervasive metaphor for artificial intelligence; and, most recently, the story has been used to caution against advances in the emerging field of synthetic biology.

This vision of Victor Frankenstein's creature as a symbol for scientific progress that comes at great cost is rooted in the novel's subtitle: *The Modern Prometheus*. Prometheus, a Greek deity, brought fire from the gods to humans. If Victor is the modern Prometheus, is the creature analogous to the fire Prometheus supplied to humankind? Fire, both an innovation and a metaphor for knowledge, was kept from humans by Zeus after Prometheus helped them trick Zeus into accepting subpar sacrificial animal parts. But Prometheus takes it upon himself to deliver this technology to humans. He pays dearly for this decision: Zeus forces him to spend eternity having his liver eaten away by eagles each day and to regenerate each night.[3] Reading the Prometheus myth as a tale of technological advancement makes sense, especially in the context of an

---

[3] Prometheus's gift from Olympus comes with a price for the god and for mankind alike. Prometheus's punishment was vivid and gruesome, but Zeus's punishment for man was Pandora and her jar of troubles. The similarities between Pandora and Eve are inescapable: women are delivered to man along with pain and suffering in retribution for newly acquired knowledge.

innovation-obsessed culture. Victor, our modern Prometheus, describes his obsession with his experiment:

> *I knew well, therefore, what would be my father's feelings; but I could not tear my thoughts from my employment, loathsome in itself, but which had taken an irresistible hold of my imagination. I wished, as it were, to procrastinate all that related to my feelings of affection until the great object, which swallowed up every habit of my nature, should be completed.*[4]

Victor's inability to focus on anything or anyone else calls to mind the mad scientist in his lab, a common image in modern Frankenstein lore as well as a popular stereotype for scientists.[5] His crime is harnessing the power of electricity to bring the creature to life. His punishment is his anguish as the creature terrorizes him and kills those he loves. He is tormented by loss and by his own inability to take meaningful action to stop the creature, a metaphoric eating of his liver each day. This price for harnessing natural forces beyond control and comprehension suggests that Victor should not have "played with fire" and that the use of electricity to ignite the spark of life was Victor's crime.

Even if we suppose that the creature is a combination of knowledge and artifact, and that Victor is the protagonist of the story, there is another interpretation: it was not

---

[4] Shelley, 82.

[5] A recent set of studies suggests U.S. residents respect science but associate scientists with immoral behavior. The stereotype of the scientist as madman seems to be alive and well, along with modern fears of Frankensteins, Fausts, and Jekylls running amok in universities and laboratories around the country. See Bastiaan T. Rutjens & Steven J. Heine, "The Immoral Landscape? Scientists Are Associated with Violations of Morality," *PLOS One*, 11, no. 4 (2016): e0152798.

the "spark" that loosed evil on the world, but Victor's refusal to guide the creature once it was alive. Victor's creation wakens to existence without malice or murder. He is not pre-wired for violence. Rather, Victor's choice to flee in disgust, and to continue to run until he could no longer escape, left the creature without guidance or further development. This new technology was not dangerous until it was abandoned. And even then, it sought oversight. To create an artificially intelligent robot might not be a crime of innovation. But to do so without embracing the responsibility for actively seeking ethical ways to use and protect the robot might be.

## Creature as Monster

*Oh! no mortal could support the horror of that countenance. A mummy again endued with animation could not be so hideous as that wretch. I had gazed on him while unfinished he was ugly then; but when those muscles and joints were rendered capable of motion, it became a thing such as even Dante could not have conceived.*[6]

Viewing Frankenstein's creation as a monster or a demon is an act of "othering" that might seem familiar to contemporary readers. Strangers who encounter the creature (mis)take him for a monster because his appearance is abnormal, even grotesque. Even Victor beholds him in terror, from the moment he first sees "the dull yellow eye of the creature open."[7] Without a word, Victor runs away and locks himself in his bedroom, as though his departure will erase the creature from existence.

There is a moment when the creature, after repeated failures to be human, makes himself into a monster. His demand for a companion created the same way he was

---

[6] Shelley, 84.

[7] Shelley, 84.

made springs from his realization that he can never belong to the human race. His morality, which until now has revealed itself through guilt over his actions and through his failed attempts to restrain himself, becomes compromised.

For Victor, choosing to see his creation as a monster is a practice of abdication: he is removing himself from his own responsibility for the creature, for the creature's dire, isolated situation, and for the creature's subsequent crimes and violent acts. While Victor may find himself awash in self-pity and shame, he does not feel a need to admit his part in the monster's creation or to warn others of the danger that approaches.

There is another way to see the creature as something other than human: as a monster of sorts, but one that can be guided or controlled. In Jewish mythology, the Golem, a beast made of clay and water, has no moral compass of its own, but can be guided by humans.[8] As rabbi and scholar of Judaica Byron Sherwin notes, the lesson of the Golem is that "creativity entails risk, but does not—as in *Frankenstein* or in the views of some contemporary philosophers, ethicists, and scientists—entail inevitable catastrophe."[9] Victor's creature might have been controlled, like the Golem, and guided toward using his strength and intellect to benefit humanity (though, as I'll discuss in the next section, this is also problematic).

If we recognize that Victor's neglect, followed by rejection from human social life, guided the creature toward

---

[8] In their introductory book series, Collins and Pinch describe science as a Golem. Harry Collins & Trevor Pinch, *The Golem: What You Should Know About Science*, 2nd ed. (Cambridge, UK: Cambridge University Press, 2012).

[9] Byron L. Sherwin, "Golems in the Biotech Century," *Zygon* 42, no. 1 (2007): 138.

his brutal behavior, we might be able to see the creature as more like the Golem than like a feral or evil creature that cannot be controlled. This interpretation leads us to the same conclusion we reached in the previous section about the creature as a technology: it is abandonment or lack of guidance that turns the creature into a monster and puts him on a path of destruction. Once again, Victor ignores an opportunity to shape the monster's actions.

## Creature as Human

*Believe me, Frankenstein: I was benevolent; my soul glowed with love and humanity: but am I not alone, miserably alone? You, my creator, abhor me; what hope can I gather from your fellow-creatures, who owe me nothing? they spurn and hate me.*[10]

Monsters are creatures of instinct, who presumably seek violence to sate some kind of base need. Humans make decisions about their actions based on the situations in which they find themselves, but because they are self-aware, they also respond emotionally and intellectually to the impact other humans' decisions have on them. Just as these discussions of the creature as a technological innovation and as a monster shed light on our beliefs about creativity, agency, and control, an examination of the creature as a human being can tell us volumes about our understanding of humanity.

The creature-as-technology and creature-as-monster interpretations omit the creature's own reflections about his condition. His longing to be part of humanity is utterly human, and his rejection, which is primarily based in his "otherness," is not so different from instances throughout history when marginalized people have sought and been denied acceptance. His perspective on what it means to be

[10] Shelley, 119.

human is often more thoughtful and empathetic than the insights offered by the human characters. He reflects on the nature of empathy when he tells Victor how he felt about De Lacey, a blind peasant the creature tries to befriend, and his family: "The gentle manners and beauty of the cottagers greatly endeared them to me: when they were unhappy, I felt depressed; when they rejoiced, I sympathised in their joys."[11] The creature's thoughtful gifts to the family, like the wood and food he leaves at their door, are heartfelt. He shows a deep empathy through his understanding that his own efforts to provide wood for the family, for example, are much easier for him than they are for Felix, De Lacey's son.

In addition to human emotions and the capacity for empathy, the creature also possesses human curiosity and a desire to explore the world intellectually, which drives him to seek opportunities to learn to speak, read, and write. He wakes without language or any comprehension of his circumstances. Immediately abandoned by his creator, he endeavors to orient himself and to understand the world around him. His ability to learn language, reading, and writing while remaining hidden from view and eavesdropping on the De Laceys is remarkable. It's true that Victor purposefully endowed him with great intelligence, but the desire and drive to pursue knowledge seems to be connected to his emotional capacity; it goes beyond the ability granted to him by his creator.

This combination of emotion and intellect allows the creature to develop his reflective nature. Each time he encounters someone who turns away from him, from his first encounter with Victor to his encounter with the small child who turns out to be Victor's young brother, he learns about the cruelty of humans. But when he witnesses the way they behave with one another, primarily through his

---

[11] Shelley, 129.

experiences with the De Lacey family, he learns about the human capacity for good and kindness. In both kinds of encounters, he takes opportunities to reflect on his situation, on the nature of humanity, and on his desire to be part of humanity. He is able to make profound observations about the way humans relate to and care for each other.

The hatred that people feel for the creature is a familiar kind of hatred that humans feel for other humans—a hatred often accompanied by or displayed through a process of "othering" that renders a fellow human so foreign that she or he ceases to be human at all. This kind of "othering" underlies the treatment of Syrian refugees in 2016, and it is reminiscent of the fear that has led to the deaths of so many unarmed Black American men, often at the hands of police officers.

The creature's revolt against his creator is not inevitable; nothing inherent in the creature's body has determined his violent behavior. It might be fair to say that the creature is a product of his own rejection. While the creature is sentient and responsible for his actions, he is not responsible for the circumstances in which he finds himself: alone in a world that will not make a place for him. His downfall comes not from any of the parts or materials Victor used to create him, nor from the spark of electricity that breathed life into him. His downfall is rooted in his most human impulse: his longing for companionship. His realization that this longing will never be satisfied ultimately compels him to abandon hope and to turn away from humanity.

After countless cruel rejections and his own murderous spree, the creature is no longer able to see himself as human. His kidnapping of Victor and demands for a companion created in his own image might seem to be an admission that he is not a human. Of course, he has been pushed to see himself as something other than—less than—human with each encounter. His violent behavior might also be seen as acquiescence to segregation. When Victor fails to deliver his bride, the creature learns to see himself as a monster. But this version of a monster actually seems to be quite human, too: his vengeance is no longer primal and driven solely by fits of rage. Instead, it is calculated. He carries out his promise to rob Victor of love the same way he has been robbed. This kind of lashing out and seeking to hurt someone who has hurt you is one of the more common strands of human frailty. He has finally accepted the role he has been taught to play by literally every human he has encountered. At last, having realized their vision of him, he can no longer bear to live.

Thinking of the creature as a human provides us with an opportunity to reconsider the interpretation of Shelley's titular assertion that Victor is a modern Prometheus. Though many think of Prometheus' gift of fire as the most notable aspect of his story, in Ovid's account of Prometheus in *Metamorphoses*, his most notable feat was not his role as the bringer of fire, but his role as creator of man. Specifically, Prometheus creates humans from clay and water, and he creates them in the image of the gods (in his own image). Because Mary Shelley's father, William Godwin, wrote an account of Greek mythology that emphasizes Prometheus's role as the creator of humans,[12] and because she spent time reading Ovid the previous year, this is likely the Promethean role foremost in her mind. [13] If Victor, as the modern Prometheus, is the modern father of a new species created in his own image, rather than the bringer of a new technology, we are forced to see the creature as human. We are also forced to look

[12] William Godwin, *The Pantheon: Or, Ancient History of the Gods of Greece and Rome* (London, UK: T. Hodgkins, 1806).

[13] Anne K. Mellor, *Mary Shelley: Her Life, Her Fiction, Her Monsters*, (New York, NY: Routledge, 1989).

more closely at the ways humans who are "othered" in some way have been treated.

## Learning from *Frankenstein*

Each of these readings of Victor Frankenstein's creature yield different interpretations of Shelley's work, with different implications. We might initially see the creature as a new and dangerous technology, but as the story unfolds, and as we read more about the missed opportunities for human connection and the creature's utter longing for companionship, we may grow to see him first as a monster with the will to destroy, and finally as the human cast out of society, who turns to brutality and revenge because nothing else is available to him. It becomes clear by the end of the story that Victor's creation is neither monster nor technology. He is utterly human, with human strengths and flaws.

Not only can we see the creature for all of his human failings, we can also see the ways that he understands his humanity, and his role in human society, in a way neither Victor nor any of the other humans he encounters can. W.E.B. Du Bois described the "double consciousness" of being Black in the United States as "a peculiar sensation, this double-consciousness, this sense of always looking at one's self through the eyes of others, of measuring one's soul by the tape of a world that looks on in amused contempt and pity." Du Bois writes of a longing to merge this double consciousness, and he imagines a world in which a Black man could live his life "without having the doors of opportunity closed roughly in his face."[14]

The creature's rejection and marginalization imbues him with this same kind of double consciousness. He

---

[14] W. E. B. Du Bois, *The Souls of Black Folk: Essays and Sketches* (Chicago, IL: A. C. McClurg & Co., 1903).

must see himself as he believes he is: full of potential humanity. At the same time, he must see himself as others see him: as something unfamiliar and terrifying. These two versions of himself are irreconcilable and painful for him, but they also give him access to knowledge about himself and those around him that they do not have.

Beyond the creature's awareness of how humans see him, he is also uniquely aware of how humans see one another. His otherness gives him a place from which to observe and thoughtfully consider human relationships. From this vantage point, the creature develops insight into the nature of humanity that the humans he encounters do not possess. Standpoint theory describes the way that people who are marginalized in society can see more clearly what those in power fail to see.[15] The creature's sustained ability to deeply consider and describe the human condition as well as his own condition represent an almost archetypal vision of standpoint theory. Perhaps he sees humanity a bit more clearly than the rest of us because he looks in on it from the outside. Perhaps he understands it more intuitively and empathetically because he is forced to consider the feelings of others as they stare at him in horror. He must see himself through their eyes, though they are never forced to consider themselves or their actions from any other perspective.

If we are to take Shelley's novel as a parable for the modern human condition, we might find that she teaches us to think more deeply about our role in shepherding invention or caring for our creations. But ultimately, she may be teaching us to look past "otherness" and connect with the humanity of the people around us. What future might the creature have had if even just one of the hu-

---

[15] See Sandra G. Harding, *The Feminist Standpoint Theory Reader: Intellectual and Political Controversies* (New York, NY: Routledge, 2003).

mans he encountered had been able to stand outside his or her own life to see the world as the creature did?

# 4

## VICTOR FRANKENSTEIN'S THREE BIG MISTAKES

## Charlotte Gordon

In 2015, Elon Musk announced the creation of OpenAI, a nonprofit artificial intelligence (AI) research company. Musk said the company was intended "to build safe AI, and ensure that AI's benefits are as widely and evenly distributed as possible." As the name suggests, the founding principle of OpenAI is essentially democratic: to make the findings of artificial intelligence researchers available to all.

Detractors believe it is unwise to allow open access to AI research. What will happen if this research gets into the hands of a "Dr. Evil," a tyrant, a fanatic, or a lunatic? Others worry that AI will surpass its human handlers and "turn itself loose on the world."

But both AI researchers and those who worry about AI should look not to *Terminator* for guidance. Instead, they should read another classic work of science fiction: Mary Shelley's *Frankenstein*. The story of a scientist's ill-fated invention of a self-directing, artificial human being demonstrates that the best protection against an evil scientist is a good scientist, and the best way to solve problems is to invite the advice of other researchers. In Shelley's tale, Victor Frankenstein, the brilliant but shortsighted scientist, made three key mistakes that could easily have been prevented by an organization like OpenAI.

**1. Isolation**: One of Victor Frankenstein's gravest errors was keeping his research a secret from others. He worked in isolation, hiding his progress from his teacher and his fellow scientists. Thus, when his creature went on a murderous rampage, killing all of those close to him, there was no one to help Victor destroy the creature or, at the very least, modify the creature's behavior. When crisis struck, there was no one to whom Victor could turn for guidance. And when Victor died, his creature continued to roam the earth, enraged and embittered, poised to wreak more damage. If Victor had been a member of a research group, his fellow scientists could have stepped in to help control the creature and to support Victor in the challenges that came to light the moment the creature attained autonomy. As it was, Victor failed to manage his invention and succumbed to the perils of the isolated researcher. He died of exhaustion and despair—a tragedy that could have been prevented by a group like OpenAI, which encourages scientists not only to share their findings, but to draw support from one another.

**2. Neglecting his creation:** When Frankenstein first beheld his creation, he was overwhelmed with remorse and disgust. He fled from its presence, giving up the opportunity to supervise, nurture, and educate his invention. In today's terms, these practices are known as "reinforcement learning" and "scalable oversight," but in essence, they add up to the same scientific principle: the inventor must carefully observe, train, and oversee his or her invention. This should have been particularly important for Victor, who had designed his creature to be as human as possible, with the supreme objective of finding love and companionship. When the creature woke to his new life and found himself alone, he experienced this as a crushing blow, a nearly fatal rejection on the part of his "father," and immediately set forth to find Victor.

As Victor regarded love as a purely positive goal, beneficial for both human beings and his creature, it did not occur to him to include a fail-safe mechanism, such as a shut-off switch. Rather, he was proud that he had designed his creation to be a free-standing and self-propelled organism. Nor did he properly consider what today's scientists would term his creature's "reward function" — the pursuit of a goal, no matter the side effects. For example, a robot with a reward function of moving an object from point A to point B will often break or destroy objects in its way unless it has been programmed properly. Along the same lines, the creature's single-minded pursuit of love meant the ruthless destruction of all that stood in the way of loving and being loved, including swift and bloody reprisals against those repulsed by him. When the creature met innocent villagers who were terrified by his appearance and took action to defend themselves, he murdered them. He burned down a family's house because they repelled his advances.

**3. Poor preparation of society and inadequate funding**: Society reacted to Victor's creation with fear and hatred — an environmental obstacle that prevented the creature from achieving its goal of love with tragic results for all involved. Due to a lack of funding, Victor had to rely on substandard materials in the manufacture of his creature. Under the cover of darkness, he dug up graves and stole the body parts of corpses. Unable to find a single dead body that had not at least partially decayed, he had to use parts from different corpses. As a result, the creature's limbs did not match and its legs and arms were in different stages of decomposition, which did not create a pleasing aesthetic. Also, the creature gave off a strong odor of decay.

However, perhaps the central reason Victor Frankenstein's creature was greeted with such antipathy was that he too closely resembled a human being—an engineering feat with unhappy consequences. Today, we would call this the "uncanny valley" effect, when an artificial agent looks too human.[1] Zombies and androids are regarded as menacing, while R2-D2 and C-3PO are seen as adorable precisely because of a central contradiction: they do not look human and yet they are all too human in their frailty and amusing idiosyncrasies.

Frankenstein's creature, on the other hand, was ugly, smelly, and frighteningly strong. His head was somewhat square, thanks to the poor quality of the skull the inventor was forced to use. This was a sad eventuality for the creature, who, on many occasions, wept about the profound revulsion others felt for him. The creature begged Victor to create another such monster, threatening him with the destruction of the human race. Without anyone to consult for guidance, Victor began construction, but before he completed the second creature, he destroyed her out of fear that she would reproduce with the first creature and annihilate humanity. In response, the creature murdered Victor's own bride, and the creation and creator became locked in a struggle that ultimately cost Frankenstein his life.

Which brings us back to artificial intelligence and, more specifically, OpenAI. Such a highly funded endeavor possesses the resources to help scientists design their creations with optimal materials and design features. It encourages researchers to work together, not in competition or in isolation, so they might advise one another. And

[1] Masahiro Mori, "The Uncanny Valley," trans. Karl F. MacDorman & Norri Kageki, *IEEE Robotics & Automation Magazine* 19, no. 2 (2012): 98-100.

when something goes wrong, a brain trust can work together to solve it.

Despite Mary Shelley's limited scientific expertise, she anticipated many of the challenges AI faces today. Few engineers can understand or predict the results of their creations. For instance, Microsoft designed a chatbot, an AI system called Tay.ai, "to entertain and engage" 18- to 24-year-olds. But in less than 24 hours, hostile users turned Tay into a troll, spouting racist, misogynistic, and anti-Semitic remarks, an experiment in AI gone awry.[2] Artificial intelligence isn't likely to kill us all—but the more people work on the problem, the more the odds go down. Victor Frankenstein's creature did not have to be a blight on society. He devolved into a monster of revenge because he was abandoned by his creator.

---

[2] Jing Cao, "Microsoft Takes AI Bot 'Tay' Offline After Offensive Remarks," *Bloomberg* (24 March 2016).

# 5

## CREATIVITY, SCIENCE, AND THE LEGACY OF VICTOR FRANKENSTEIN

## Sara Imari Walker

We humans are a creative species. We create art. We create satellites to launch into space. We create fantastic works of fiction about realities that have never existed. In Mary Shelley's *Frankenstein*, we also create life—with monstrous consequences. After a gruesome and painstaking experiment executed over many months, young Victor Frankenstein is finally successful in bringing inanimate matter to life, only to then face a cascade of tragedies. In retelling of his own story, Victor reveals a devastating series of events stemming from his act of creation. There has been much debate about the genesis of the tragedy in *Frankenstein*: Is it Victor's hubris in assuming the role of creator? Or is it his inability or unwillingness to take responsibility for his act? Either interpretation casts a bleak light on the pursuit of scientific discovery and on our ability to act responsibly as a creative species.

Victor was investigating some of the deepest questions in science, questions that still puzzle us today: *What is life? How does it arise from inanimate matter?* Creating life from non-living matter remains an elusive scientific endeavor, even in the light of the molecular revolution of the 20th century. The horizon is constantly receding: every few years there seems to be a new estimate that life will finally be synthesized from scratch in the *next* five to ten years. In this respect, the state of the art today is perhaps not so different from Shelley's day, when learned people also thought that the problem of understanding the mechanisms of life would soon be solved. At that time, a common scientific viewpoint was that life emerges spontaneously all the time, under everyday conditions. Consider a loaf of bread left out on the kitchen counter for several weeks. Life seems to spontaneously emerge on its surface in the form of common mold — often despite our best efforts to prevent it. So it's not surprising that we once thought that life must form readily from non-living matter, and therefore that it should be relatively easy to create life using basic scientific principles.

In 1828, when Fredrich Wöhler demonstrated the synthesis of urea (an organic molecule) from cyanic acid and ammonium, the idea of creating life in the lab became a serious scientific endeavor. This was just a few years after the first edition of *Frankenstein* was published in 1818, so Shelley certainly in some sense anticipated science just on the horizon. Wöhler's experiment took down the idea that producing the stuff of life required a vital force, or *elàn vital*, by revealing for the first time that organic molecules could be produced from non-living ingredients. However, it was not until Louis Pasteur's famous 1859 experiment demonstrating that meat broth does not spoil if appropriately secured from contamination that the idea of "spontaneous generation" (in the form of continuous, spontaneous formation of fully formed organisms) was finally disproved. The mold that springs up on your bread after a few days or weeks does not arise spontaneously, but rather descends from a successful evolutionary lineage of mold cells that found a happy place to set up shop and proliferate.

The problem of creating life is much more difficult than one might expect. The origin of life on Earth (thus far our only example of inanimate matter becoming animated) is a remote event that happened more than 3.5 billion years ago, and the details have been lost to evolutionary history. The problem of creating life is made more difficult still by the fact that we know now — thanks to the molecular revolution spurred by the discovery of the DNA double helix — that we have only one sample of life with which to understand it. Tracing the evolutionary history of life on Earth reveals a common ancestry, suggesting that life may have arisen only once on our planet. We are the direct product of that event, after several billion years of evolution. This makes it difficult to know how life got started. We can't even say with certainty if the first living systems shared the same chemistry that life uses today. We have yet to identify what, if any, universal features might explain life and therefore allow us to reconstruct its origins, and also to synthesize genuinely *new* life in the lab in a real-world Frankenstein experiment (but most likely at the microbial level).

Because of these constraints, and the simple fact that we don't really know what life is, all attempts so far to create life from scratch have involved taking small, systematic steps outward from known biology. An example is the genome transplant performed at the J. Craig Venter Institute, which demonstrated the transformation of one bacterial species into another by the transplantation of a complete genome. In contrast to Victor's experiment, here the subjects were microbial organisms, rather than human body parts, and they remained animated through the duration of the experiment — no postmortem reanimation was involved. The experiment revealed that the algorithm encoded in the transplanted DNA could completely rewrite the chemical "hardware" of the recipient cell, transforming both genotype and phenotype to that of the donor organism. Basically, the donor cell was overwritten to become the same species as the recipient cell in terms of *all* of its cellular machinery—not just the DNA. Instead of demonstrating how to create brand-new life (as it has sometimes been described in popular media), what the experiment demonstrated is that the "software" of life—a cell's genomic information content—can be transplanted between related organisms. Remarkably, once the software is transplanted, the chemical hardware will in turn know how to interpret it to produce the right machinery.

The analogy to computing here is significant. Just as many software programs will run on a variety of computers, it seems that the software of living cells can be transplanted between organisms and still be operational. However, in this example the actual operating system (OS) was the same to begin with—that is, the two cells represented the same tree of life united by common ancestry. In fact, the two organisms were closely related species. It remains an open question whether a similar experiment would be successful in transplanting a genome between two distantly related organisms. And it's another matter entirely whether we could even in principle boot-up life with a completely different OS, or reboot a system by rewiring, like Victor did when he assembled his creature from cadaver parts.

However, there is some reason for hope that we might one day accomplish such a feat. In addition to the Venter Institute's genomic transplant, an increasing number of examples illustrate that the hardware of life (its chemistry) is malleable, so long as the software (i.e., genomic information) is preserved. Synthetic biology is now moving at a rapid pace. Alternative genetic molecules (other than RNA or DNA) and genomes with expanded alphabets of more than four nucleobases (in addition to A, G, C, and T as found in natural DNA) are now being demonstrated to be viable within living cells. These examples demonstrate how far we have come in manipulating the biochemical structure, or hardware, of known life. But this is still far removed from Victor's accomplishment. To genuinely create life from scratch requires knowing the principles by which life operates, and these have yet to be uncovered by science.

We can, however, glean clues about these principles from the analogy of describing life in terms of its chemical hardware and informational software. Increasingly, life is being described in terms of information, computation, and even learning. Take, for example, a typical description of the interior operation of a cell, which often utilizes informational language such as "transcription," "translation," "codes," "sensing," and "signaling." It may not be a coincidence that an effective way to describe the genomic transplant performed at the Venter Institute is in terms of booting up software within the hardware of the host cell. For example, the single-celled slime mold that can also sometimes act as a multicellular aggregate will learn, when given a pattern of food sources matching (to scale) the positions of metro stations in Tokyo, to generate the same network solution for accessing those food resources as the metro engineers designed for moving people.[1] We could describe even evolution itself in terms of what a species has learned through evolutionary history: we might understand a genome as a repository of information about an organism's environment.

In *Frankenstein*, Shelley never divulges a working definition of life, and we never learn about Victor's insights into animate matter that allow him to create life. However, in light of the informational perspective on life emerging in the 20th century, we might conclude that what

---

[1] Atsushi Tero, Seiji Takagi, Tetsu Saigusa, Kentaro Ito, Dan P. Bebber, Mark D. Fricker, Kenji Yumiki, Ryo Kobayashi, and Toshiyuki Nakagaki, "Rules for Biologically Inspired Adaptive Network Design," *Science* 327, no. 5964 (2010): 439-442.

Victor really did was discover the assembly language of life and use it to write an entirely new OS to animate the creature.

How we arrive at the foregoing computational view of biology—or any other future scientific answer to the question *What is life?*—is by actively questioning what we know and learning from our discoveries and the things we create. That is, by doing science. If Victor had looked more closely when he was unraveling the principles of life, he might have realized that much of the biological imperative is precisely to acquire information about one's environment—that is, to learn. In a very physical sense, creativity may be tied to learning, because it's through a process of learning about the environment and adapting accordingly that natural selection has created such diversity of life on Earth. A vivid example is that of niche construction, where adaptation leads organisms to alter their environment (a creative act, by recreating the environment in which they live) and in turn, they must adapt (learn) from their creations or face extinction.

Victor, in contrast, did not have the goal of learning in his act of creation. Thus his was an act of creation decoupled from a process of learning—very unlike life. He did not care to understand the creature once he had created him. In his hubris, Victor assumed he knew everything and had nothing to learn. Yet surely there was much to be learned from the creature, perhaps in the process avoiding the great tragedy that was Victor's fate. Initially, Victor may have embarked on his quest out of curiosity, but it quickly transformed into an act of ego. Once the creature was alive, Victor became ruled by fear of what he had created rather than curiosity. This is the tragedy of *Frankenstein* and marks most of the stigma surrounding the scientific endeavor. But this is not science. Science, perhaps much like life itself, is a creative act. How we treat our creations determines their impact, and that may or may not be encompassed within the realm of science. The role of the scientist is to learn, it is up to us as a species if we let our creations rule us or if we instead learn from our creations. Almost a century after Shelley's creative act of writing *Frankenstein*, another famous female intellectual, two-time Nobel Prize winner Marie Curie, eloquently stated what perhaps should have been Victor's perspective, if he had continued on the path of the scientist instead of the path of fear: "Nothing in life is to be feared, it is only to be understood. Now is the time to understand more, so that we may fear less."

This is perhaps the deepest tragedy of Victor Frankenstein and his legacy. Not his hubris, nor his inability to take responsibility for his actions, but rather that *he stopped being a scientist* at the moment his creation came to life. By so doing, he lost not just his humanity but something intrinsic to our very nature as living systems: our desire to learn.

# 6

## WHAT VICTOR FRANKENSTEIN GOT WRONG

### Kevin Esvelt

It seems banal to state that the future of our civilization will be determined by the technologies we invent and the wisdom with which we deploy them. That banality may explain why we collectively spend so little time concerned with how best to proceed. Literature offers many cautionary tales, but we seldom pause to re-evaluate them in light of modern capabilities. Early in Mary Shelley's *Frankenstein*, Victor Frankenstein issues a personal warning:

> *Learn from me, if not by my precepts, at least by my example, how dangerous is the acquirement of knowledge and how much happier that man is who believes his native town to be the world, than he who aspires to become greater than his nature will allow.*[1]

We are fortunate beyond measure that so many great scientists ignored this advice. From vaccines and antibiot-

---

[1] Mary Shelley, *Frankenstein; or The Modern Prometheus*, ed. D. L. Macdonald & Kathleen Scherf, 3rd Ed. (Peterborough, Ontario: Broadview Press, (2012) [1818]), 80.

ics to abundant food and energy, technological advances have liberated most of humanity from the worst of disease and want. Our gains may be fragile, requiring a steady diet of new discoveries to maintain, yet they are real and deserve celebration.

Still, it is hubris to assume that past triumphs guarantee any measure of future success. As a research scientist currently working on a controversial new approach to ecological engineering,[2] I recently reread *Frankenstein* and was surprised by its depth. Far from the popular characterization of "scientist meddles with life, tragedy ensues," Victor Frankenstein is a deep yet flawed human being whose mistakes are relevant for researchers today. Whether or not it was Shelley's intent, the novel's message is clear: wisdom is knowing whether, when, and how to develop new technologies — and when to lock them away for as long as we can.

In 2013, after helping to develop a gene editing tool called CRISPR — a molecular scalpel for precisely cutting and therefore editing any DNA sequence — I realized that we could use it to construct "gene drive" systems to alter the traits of wild populations. Here's how it works: instead of just using CRISPR as a tool to edit DNA once, we can program the organism's genome to do the editing on its own, then let it mate with a wild counterpart. In the offspring, CRISPR will convert the original DNA sequence inherited from the wild parent to the new, edited version. With two copies, the next generation is guaranteed to inherit the edit — as is the next, and the next, and the next. Think of it as a find-and-replace for entire wild populations. The implications could be profound: as Austin Burt of Imperial College London first predicted more

---

[2] Michael Specter, "Rewriting the Code of Life," *New Yorker* (2 Jan. 2017).

than a decade ago, learning to harness gene drive could let us stop mosquitoes from spreading malaria.[3]

No prose can do justice to the sheer elation of discovery, though Shelley makes a worthy attempt. As Victor Frankenstein relates, "I trod heaven in my thoughts, now exulting in my powers, now burning with the idea of their effects." Such was my experience when I realized the implications of merging CRISPR with gene drive to create a tool that could help eradicate diseases, save endangered species, and obviate the need for pesticides—a way to solve ecological problems with biology, not bulldozers.

But the thrill of invention can amount to a siren song. As Robert Oppenheimer famously said, "When you see something that is technically sweet, you go ahead and do it, and you argue about what to do about it only after you have had your technical success. That is the way it was with the atomic bomb."[4]

It need not be so.

Most technologies require substantial resources to deploy and impact the world—think of the steam engine, electricity, and vaccines. Those few that do not, such as software, still typically require voluntary adoption by many other people. These constraints normally give individuals an opportunity to selectively opt out, as do the Amish, and more broadly provide society with a chance to consider the ramifications.

With CRISPR-based gene drive, however, anyone with the right training could conceivably alter whole ecosystems unless his or her creation is actively countered and

---

[3] Austin Burt, "Site-specific selfish genes as tools for the control and genetic engineering of natural populations," *Proceedings of the Royal Society B: Biological Sciences* 270, no. 1518 (2003): 921-928.

[4] Quoted in Richard Rhodes, *Dark Sun: The Making of the Hydrogen Bomb* (New York, NY: Simon & Schuster, 1995), 476.

overwritten. In the worst-case scenario, an unopposed global gene drive system could spread through every population of the target species in the world, potentially affecting countless people without their consent. While most genetic changes would have no ecological effects whatsoever, we can't know for sure without testing them in small areas (i.e., without a global gene drive), and individuals acting on their own won't have run such tests. Imagine if someone in, say, New Zealand — even a would-be do-gooder — released a global gene drive designed to suppress or remove an invasive rat population by spreading infertility.[5] Even if it worked well, possibly saving many endangered species, the construct wouldn't stay in that area. It would spread (or be spread) by ship or plane to Eurasia and likely collapse the native rat populations there, with unknown ecological consequences. (This is a major reason why gene drive researchers have emphasized laboratory safeguards,[6] and also why my group is working to develop daisy drive, a form of CRISPR-based drive system that runs out of genetic fuel and stops.)

Thankfully, few people have the necessary skills and knowledge to insert genes into organisms that reproduce sexually, and the vast majority of them work with laboratory fruit flies — hardly a keystone species. After three

---

[5] Eleanor Ainge Roy, "No more rats: New Zealand to exterminate all introduced predators," *The Guardian* (25 July 2016).

[6] Omar S. Akbari, Jugo J. Bellen, Ethan Bier, Simon L. Bullock, Austin Burt, George M. Church, Kevin R. Cook, Peter Duchek, Owain R. Edwards, Kevin M. Esvelt, Valentino M. Gantz, Kent G. Golic, Scott J. Gratz, Melissa M. Harrison, Keith R. Hayes, Anthony A. James, Thomas C. Kaufman, Juergen Knoblich, Harmit S. Malik, Kathy A. Mathews, Kate M. O'Connor-Giles, Annette L. Parks, Norbert Perrimon, Fillip Port, Steven Russell, Ryu Ueda, and Jill Wildonger, "Safeguarding gene drive experiments in the laboratory," *Science* 349, no. 6251 (2015): 927-929.

years of evaluating potential dangers, my current best assessment is that gene drive is unlikely to present much of a biosecurity threat, or even a major ecological hazard. Simply put, drive systems spread slowly, can be unfailingly and cheaply detected, and are easily overwritten and therefore countered.

But it would be sheerest hubris to assume those are the only concerns. If anyone unilaterally set in motion a process that might alter an entire wild species, even if it didn't work well or had no measurable effects, the consequences for public trust in scientists and governance could be devastating — perhaps enough to cripple research that our civilization desperately needs.

It's important to note, however, that there is a major difference between gene drive and the fictional tale of Mary Shelley: Victor Frankenstein did not ask anyone for advice.

Many of my colleagues and mentors — particularly George Church, Kenneth Oye, Jeantine Lunshof, and James P. Collins — have worked with me and others to examine gene drives and their potential consequences. Along with numerous other experts from diverse fields, including representatives from environmental organizations, we discussed the implications, risks, and benefits, ultimately concluding that it was not only safe to tell the world about our discovery, but ethically necessary to do so *before* anyone tested it in the laboratory.[7]

Had it been up to my judgment alone, things may have gone as badly as in fiction.

---

[7] Kelly Drinkwater, Todd Kuiken, Shlomiya Lightfoot, Julie McNamara, and Kenneth Oye, *Creating a Research Agenda for the Ecological Implications of Synthetic Biology* (Washington, DC: Wilson Center, 2014).

Even fiction could have been worse. If Victor Frankenstein had created a fertile mate for his creature, it would surely have represented an existential threat to humanity — likely the first such technological example in literature. Had he shared the secret with others, someone else would surely have done something equivalent. (When asked about "the particulars of his creature's formation," Victor's response is, "Are you mad, my friend?"[8]) Yet with that same technology, humanity might also have abolished disease and aging, hunger and want, perhaps rendering us invulnerable to that type of existential risk. Evidently the fictional inventor didn't think of that. And even if he had, his lone evaluation of the risks and benefits would not be nearly as accurate as if he had consulted with a diverse group.

Technological hubris is ignoring the suggestions of others — even if only by neglecting to inform them of an advance. It is most common among those suffering from the curse of knowledge: scientists.

That's why my colleagues and I seek to ensure that all gene drive research takes place in the open light of day.[9] People deserve a voice in decisions that might affect them, and building gene drive systems behind closed doors denies them that opportunity. Even apart from the moral hazard, keeping research plans secret — as the current scientific enterprise incentivizes us to do — is appallingly inefficient and outright dangerous. It doesn't just slow the rate of advances, thereby jeopardizing our ability to sustain our civilization; it practically invites global catastrophic risk.

No one, be they a science fiction author or Austin Burt himself, anticipated a form of gene drive as versatile as is

---

[8] Shelley, 209.

[9] Kevin Esvelt, "Gene editing can drive science to openness," *Nature* 534, no. 7606 (2016): doi:10.1038/534153a.

theoretically enabled by CRISPR. What else have we not anticipated that this time might be truly dangerous? And given this possibility, why on earth do we send out small teams of ultra-specialists, mostly working on their own and in secret, to find and open every technological box they can? Better to default to open research plans, enabling diverse teams to evaluate new advances, implementing measures to obscure and counter anything deemed truly dangerous, than to proceed blindly.

Of course, any wholesale restructuring of the scientific enterprise would also be an act of reckless hubris. My personal rule of ecological engineering: start local and scale up only if warranted. In this case, the best "local test" is the field of gene drive research. Scientific journals, funders, policymakers, and intellectual property holders should change the incentives to ensure that all proposed gene drive experiments are open and responsive.

The message from fiction and reality is clear: scientists should hold themselves morally responsible for all consequences of their work. The least we can do is muster enough humility to ask for help.

# 7

## THE INSUFFICIENCY OF COOL

## David Guston

My son Sam was stomping through a springtime puddle when he was three and a half. With less than eight inches of rain in our average Arizona year, I indulge him his splashing. In an impish mood myself, I admonished him: "Watch out for puddle gators!"

He looked up, slightly startled, paused, and replied, "Daddy, there is no such thing as puddle gators."

"Wouldn't it be cool if there were?"

"YES!" He added a splashing boot-stomp for emphasis.

As Sam and I speculated on the natural history of our imaginary puddle gators, I was nevertheless aware of the divide between my imagination and the actual ability of anyone literally to create such a wee beastie.

But I am deeply ambivalent about the research agenda of synthetic biology, a field of research that could, eventually, make puddle gators a reality. Synthetic biology sees organisms as objects to be engineered. It sees biological parts — proteins, the metabolic pathways and systems that produce them, and the genes that encode them — as building blocks that are interchangeable among organisms. Plug some salamander traits like size and coloring into the appropriate "chassis" of a clone-able amphibian, add some alligator teeth, and you've got ... a puddle gator!

Puddle gators could be only playthings or ecological intrusions. But creating a male *Aedes aegyptus* mosquito — the vector for dengue fever, Zika, chikungunya, and yellow fever — with genes that would produce only unviable offspring could save untold thousands of lives.

Doing stuff like engineering puddle gators because it's cool — in a carefully controlled environment with plenty of contextual instruction — can help encourage creativity and develop capabilities to address larger problems, such as disease. But fostering curiosity should not foster hubris. Curiosity is not a reason for endangering others or the environment.

☷    ☷    ☷

Two hundred years ago, in the cold, wet June of 1816, the so-called Year Without a Summer, a young woman responded to a dare among friends to write the scariest thing each could imagine. Eighteen years old, she was holed up with them in a villa on the shores of Switzerland's Lake Geneva. They were awed by the Alpine echoes of massive lightning storms, fueled by ash from the eruption of Mt. Tambora, half a world away in present-day Indonesia, the previous year. Probably doing laudanum — opium mixed with alcohol — and spending intimate time with a man married to another woman, the young woman is of course Mary Wollstonecraft Godwin, and the married man was her still-future husband, Percy Bysshe Shelley. The others were Lord Byron, Byron's friend and doctor John Polidori, and Mary's stepsister Claire Clairmont. The scariest thing Mary could imagine was *Frankenstein; or, the Modern Prometheus*, a story about a young science student named Victor Frankenstein who delves into the secrets of alchemical texts and discovers how to revivify a body composed of various pieces of other, previously dead bodies — with disastrous consequences for him and his loved ones.

Before writing *Frankenstein*, Mary Shelley witnessed public demonstrations of the vital power of electricity, called down from the heavens and at least partially tamed by the likes of Benjamin Franklin, Luigi Galvani, Giovanni Aldini, and others. Such demonstrations featured the electrically induced twitching of the muscles of newly dead frogs, dogs, and other beasts — including humans — in a lifelike way.

Because of his monstrous, unnamed, electrically induced creation, the name Frankenstein represents curiosity taken to the level of hubris, rightly I think, because Victor had little else in mind than the grandiose goal of creating a more perfect being. Once he failed to create something beautiful as well as swift, strong, and smart, Victor shrank away from responsibility toward his creation. The creature was left to fend for himself and learn about human kindness and cruelty, ultimately confronting Victor with the fatal consequences of his moral failure as creator and father.

Victor has a lot in common with contemporary synthetic biologists. He understood life to be composed of interchangeable parts, and that a scientist could create life by bringing together non-living parts with the right skill and a little jolt of something (electricity for Victor; the information encoded in DNA for synthetic biologists).

Among the agendas of synthetic biology, beyond the terminator mosquito, is a project on the so-called "de-extinction" of some species of animals. In Australia, not long ago, scientists nearly succeeded in de-extincting the gastric-brooding frog. This "Lazarus Project" also has its sights set on de-extincting the Tasmanian tiger.[1] Other scientists are after the woolly mammoth, and, just past the centennial of its disappearance, the passenger pigeon, which some believe to have been the most numerous terrestrial vertebrate on the planet before it was driven to extinction through hunting and habitat destruction.[2]

De-extincting these creatures might provide some measure of dispensation for our human role in their disappearance. But motivations are complex things. Science writer Carl Zimmer, in *National Geographic* in 2013, explored some practical and ethical concerns about de-extinction. He closed his piece by quoting the Stanford University bioethicist Hank Greely. "For Greely, as for many others," Zimmer wrote, "the very fact that science has advanced to the point that such a spectacular feat is possible is a compelling reason to embrace de-extinction, not to shun it. 'What intrigues me is just that it's really cool,' Greely says. 'A saber-toothed cat? It would be neat to see one of those.'"[3]

Some scientists have even floated the idea of de-extincting Neanderthals, as Harvard University's George Church did during a 2013 interview with *Der Spiegel*.[4] Now, I don't want to discourage scientists from discussing things in public that are not yet possible in the laboratory. But what could be more Frankensteinian than birthing into the world a unique being, a near-person with

[1] Jorge Branco, "The Lazarus Project: Scientists' quest for de-extinction," *The Sydney Morning Herald* (18 April 2015).

[2] Nathaniel Rich, "The Mammoth Cometh," *New York Times Magazine* (27 Feb. 2014).

[3] Carl Zimmer, "Bringing Them Back to Life," *National Geographic* (April 2013).

[4] "Can Neanderthals Be Brought Back from the Dead?" *Der Spiegel* (18 Jan. 2013).

no prospect of biological kinship and little hope of human sympathy?

To explore this terrain further, let's look again at the tale if the Golem that Megan Halpern recounted in Chapter 3. While it is a religious tale, it is not related to transgressions against or responsibilities to God, the way religious tales are often interpreted. As a Jewish tale, it involves a prominent element of Jewish ethics that human beings are co-creators with God. In the myth of the Golem, which pre-dates *Frankenstein* by several centuries, a man makes an animated creature from inanimate stuff — closer to the original creation, as the stuff of the Golem is clay and the spark is the name of God itself.

The man in the myth is often the Rabbi Judah Loew, a historical figure who presided at the Staronova, or Old New Synagogue in Prague, and is buried there. There are many versions of the myth, including a book by Elie Wiesel, a subplot by Marge Percy, and the German expressionist film by Paul Wegener (1920), in which he stars in the title role of *Der Golem*. Rabbi Loew is said to make the Golem as a last resort, to protect the Jews of Prague from a pogrom that everyone knows is coming, but that no one, not even the king who made use of the rabbi's learning and the Jewish community's taxes, will stop. Exceptionally strong and obedient, the Golem fulfills his purpose and saves the community from slaughter. In some versions of the story, the Golem then becomes a servant in the synagogue and in the house of Rabbi Loew, where the rabbi lives with his daughter.

The harried but also haughty rabbi neglects his paternal duties in educating and socializing the Golem, at least until the monster falls in love with the rabbi's daughter. At this point in the story, the rabbi takes advantage of part of the mystery of the Golem's creation, the Hebrew word *emet* for "truth," written across its forehead. The Rabbi erases the first letter, *aleph*, leaving the Hebrew word *met*, or death, and the Golem dies. It is a sad and even cruel moment when the Rabbi Loew erases the life of the Golem, especially because it is the rabbi's own human flaws, and not the Golem's, that precipitate the untenable situation.

Although his approach is not altogether satisfactory, compared to Victor Frankenstein the Rabbi Loew has done a bit better. He has a precise, community-regarding, and justice-serving reason for engaging in this extreme act of creation, something that Victor lacks. And while the Rabbi shares Victor's culpability as a bad father, he at least retains control over a kill switch — something that Victor never had or even contemplated.

In some instances, like the *Aedes aegyptus* mosquito, contemporary biologists have a precise, community-regarding, and justice-serving purpose in mind. And they have created a kill switch in the engineered mosquitos which, one hopes, might work should they release this creature — new to evolution — out into the world. But for the scientists involved in the de-extinction work, I worry that their purpose might be more self-regarding than community-regarding, more general than precise, and the purpose of de-extinction is intentionally antithetical to having a kill switch.

Perhaps the places where we 21st century innovators might learn to be responsible are, ironically, from the absurdity of comedy and the innocence of childhood. In Mel Brooks' *Young Frankenstein* (1974), Victor's descendant Frederick Frankenstein responds initially with now-familiar horror at his creation. But then, shouting "Hello, handsome!" in a blaze of psychological insight born of sheer terror, Frederick figuratively and then literally embraces his monster.

In Tim Burton's *Frankenweenie* (2012), preadolescent Victor discovers the secret of revivification and applies it to his beloved dog Sparky. The secret gets out, and other children go about reviving dead pets that then run amok in the traditional trope of the monster movie. The process of revivification, however, is incomplete, as Sparky does not revive until young Victor's tear drops on him. Victor has been a good and loving master to Sparky, who remains loyal and loving back.

Frederick and his monster and Victor and Sparky can show us how to love our monsters, so we don't need—as the Rabbi Loew does—to put them out of our misery.

In light of my son Sam's love of the natural world, I often lament that "natural historian" is no longer a career open to him. As with the puddle gator, however, I love the way he melds his imagination with the natural world—something that natural historians, and especially those turned into field biologists, are not often encouraged to do. Synthetic biologists, however, seem to partake in this melding of imagination and observation, and that is likely a career open to him. So I also lament the continuing shortsightedness of an innovation system that does not emphasize responsibility.

For now, while Sam is still learning the rules of social behavior along with the names of Arizona critters, such fantasy is cool. Responsibility—the ability to reflect on, articulate, and follow a precise, community-regarding, and justice-serving purpose, and to plan to stop if something should go wrong—comes later. But for the day he splices his first gene, I hope that he will have already learned that "Wouldn't it be cool if …?" is not sufficient justification. What would be really cool is if he learned that from his science teacher, and his art teacher, and a whole host of others, and not just from me.

# 8

## MODERN-DAY FRANKENSTEINS

## Alyssa Sims

Mary Shelley's *Frankenstein* is sometimes considered the first science fiction novel. It has had massive influence on the way we think about scientific research, giving rise to the idea of the mad scientist. But tragically, one important thing is often overlooked: Victor Frankenstein was a pioneer in do-it-yourself science.

The spirit of Shelley's title character is reflected in the modern movement of biohackers, a subculture of professional scientists and amateur enthusiasts who are developing community spaces for scientific exploration, discovery, and, fundamentally, learning. This garage biology movement, as it's been called,[1] exists on a spectrum. On the extreme end, some are using do-it-yourself (DIY) science to hack their own anatomies in pursuit of a sort of transhuman ideal. But the broader, mainstream DIYbio community simply seeks to encourage public engagement with science.

While the term "biohacking" wasn't tossed around until roughly the early 2000s, the inclusion of outsiders in

---

[1] William Saletan, "Faking Organisms," *Slate* (1 Feb. 2011).

knowledge production — and for that matter, the occurrence of scientific discovery outside of academia — is an old one. Rob Carlson wrote in *Wired* in 2005, "The era of garage biology is upon us,"[2] but in 2009, Stanford University bioengineering professor Drew Endy suggested to the *San Francisco Chronicle* that, in fact, it may have been upon us a long time ago: "Darwin may have been the original do-it-yourself biologist, as he didn't originally work for any institution."[3]

Still, it looks a lot different today than it did during the Romantic era, when Shelley was writing *Frankenstein.* There is a video online, for example, of a man using a remote to control his friends — it's somewhat alarming to watch.[4] Using a magnetic helmet, he was able to steer a person by tapping into his vestibular apparatus — the stuff in our ear canals that provides our sense of balance and spatial orientation. By sending electrical currents to different sides of the helmet, he is able to trick the wearer into believing that ground is in a different direction, causing the brain to overcompensate and push the wearer in the opposite direction of the currents to regain balance. Pretty cool stuff, if you're down with turning your friends into science projects.

This kind of experiment is more representative of the grind scene, a subset of the biohacking community, than the community at large. "Grinders" are to do-it-yourself biology as punks are to the rock music genre, espousing progressive — even anarchist — views about the relationship between humans and society. In a 2012 piece in the *Verge*, a biohacker named Shawn Sarver described

---

[2] Rob Carlson, "Splice It Yourself," *Wired* (1 May 2005).

[3] Julian Guthrie, "Do-it-yourself biology grows with technology," *San Francisco Chronicle* (20 Dec. 2009).

[4] https://www.sciencechannel.com/tv-shows/outrageous-acts-of-science/videos/man-uses-mind-control-on-his-friends

himself and a collaborator as "technolibertarians," anti-authority figures with a penchant for techno-scientific body augmentation.[5] In the *Verge* piece and an accompanying short documentary, the term "wearable technology" takes new meaning as Sarver allows a local piercing artist to implant a miniature magnet into the tip of his finger. The dingy, backroom operating theater, coupled with Sarver's steampunk aesthetic, creates a scene that is less sci-fi and more reminiscent of *Mad Max*. He downplays the pain of the bloody, indelicate procedure, which consists of a scalpel inserted a couple of centimeters into his digit, the lodging of a magnet, and an unprofessional suture job, all without the aid of any anesthetic. At the conclusion, Sarver appears genuinely moved by his peculiar new power: the ability to attract paper clips and other small metals to his finger from an extremely close distance. Perhaps the central conceit of the grinders' ethos is the perception that their augmentations are not simply out of curiosity of their own anatomies, but for the greater good—an evolutionary fast track that, one day, will benefit all of humankind.

But not all biohackers are actually hacking up their bodies. Far more are tinkering with DNA in ways that produce exciting outcomes and sharing their findings and expertise with anyone interested in learning. They're creating spaces to support community engagement with, and the democratization of, science. In effect, then, they are questioning the power, authority, and hierarchy of academic institutions. As Alessandro Delfanti alluded to in his article "Tweaking Genes in Your Garage," it poses a direct challenge to the monopoly power and capitalism of Big Bio, "the ensemble of big corporations, global univer-

[5] Ben Popper, "Cyborg America: Inside the Strange New World of Basement Body Hackers," *The Verge* (8 Aug. 2012).

sities, and international and government agencies that compose the economic system of current life sciences."[6]

This is where DIYbio comes in. DIYbio, which has been around since 2008, is an informal organization that helps to organize biohacking individuals and groups with similar goals. While the DIYbio movement has yet to engineer a serious breakthrough, the projects in this space range from the curious, like using commercially available toolkits to search for new antibiotics in soil samples, to the extraordinary, such as manipulating bacteria to produce insulin or even isolating cells from organs — a project that, if successful, could increase the success rate of organ transplants.

In 2011, the North American DIYbio Congress drafted the code of ethics that most biohackers ascribe to.[7] (Europe has a separate but similar code.) Perhaps most importantly, the code enshrines "tinkering" as a value. The other tenets of the code include: open access, transparency, education, safety, environment, and peaceful purposes, emphasizing their commitment to responsibility.

But DIYbio also facilitates the movement's relationship with government entities. Do-it-yourself biology has come under the scrutiny of the Federal Bureau of Investigation and the Presidential Commission for the Study of Bioethical Issues. The fear is that today, a basement-scientist terrorist could create and release a dangerous pathogen. An outbreak wouldn't even have to be intentional — when

---

[6] Alessandro Delfanti, "Tweaking Genes in Your Garage: Biohacking Between Activism and Entrepreneurship," in *Activist Media and Biopolitics: Critical Media Interventions in the Age of Biopower*, Wolfgang Sützl & Theo Hug, eds. (Innsbruck, Austria: University of Innsbruck Press, 2012), 163-177.

[7] "Draft DIY Code of Ethics from North American Congress," available online at: https://diybio.org/codes/code-of-ethics-north-america-congress-2011/.

you're tinkering with biological materials, as biohackers are, it can be hard to protect against unpredictable mutations and evolutions of bacteria in an uncontrolled environment.

And that brings us back to *Frankenstein*. The reality is that it still requires a high level of scientific expertise and sophistication to engineer a worse-case scenario — say, an airborne strain of the Ebola virus — outside of a traditional laboratory. But even the simplest project can have unintended consequences if materials and chemicals aren't handled with care. And when biohacking experiments go awry, there is no governing body or academic institution to hold someone accountable.

"I don't think Frankenstein set out with bad intentions, but there are some things that are unpredictable," said Elizabeth Tuck, a DIYbio Meetup group organizer in Washington, DC with a background in molecular biology and genetics. "The unknown unknowns are real, and if we don't acknowledge that those things are real and important, then we're being disingenuous. So I think it's important for us scientists to have a lot of humility about what we're doing. If things go badly, we have to be willing to say, 'We screwed up, and here's what that means.'"

The things that really distinguish Frankenstein from today's biohackers are their code of ethics and the emphasis on collaboration. After all, it was not his pursuit of radical science that was flawed — it was his response to failure. Instead of asking for help when he became horrified by his creation, Victor fled. "If you come in here we're not going to be, you know, 'Frankensteining' for real," said Tuck. "Because we have community ownership and community participation, I think we can actually hold ourselves to a very high standard of conduct for people to actually be thoughtful and intentional about how they are engaging with this kind of science."

# 9

## FRANKENSTEIN'S METROPOLIS

## Jathan Sadowski

When you imagine a contemporary incarnation of Victor Frankenstein, the picture that pops into your head is probably something close to a biotechnologist. Clad in a white coat and ensconced in a laboratory, the 21st century Victor deftly handles his sophisticated equipment. Perhaps applying precise amounts of electric charge to a specimen or manipulating its genetic material, he works to animate his creation. And instead of being the archetypal lone genius—a careless, brash, ambitious researcher who works madly to make his *"Eureka!"* discovery—he is likely part of a team dedicated to the advancement of innovative outcomes.

But what if Frankenstein had decided to forego his biotechnology career altogether? What if instead he decided to become an urban planner, creating "smart cities" in place of a smart creature? Rather than building creations that are atomistic and aware, he decides to construct whole smart systems of people, places, and technologies. Rather than working with limbs and cells, his medium is the urban environment coupled with information technology. Rather than Frankenstein's monster, he gives us Frankenstein's metropolis. The cities he creates are alive and crackling with energy, but thanks to data, sensors, and digital networks—not blood, organs, and galvanic vitality.

It is not hard to imagine Victor Frankenstein, who had the desire to create something intelligent, autonomous, and reactive, being lured in by the promise of technology that can process information, make decisions, and respond to its environment. It is from this premise that I use *Frankenstein*'s themes—of power, creativity, and control, for example—to illuminate some of the dynamics at play in so-called smart cities. *Frankenstein* provides a way of fleshing out the relationships among creator, creation, and the rest of us—and the social and ethical consequences of this relationship. One of the creations under discussion is obviously the city itself, with its newfound digital liveliness and capacity for sensing and responding. Less obviously, the creations are also the city inhabitants, and their bodily entanglements with smart environments.

I will begin by introducing the smart city, helping to orient us towards the ideas and practices of this movement. From there, I will explore concerns about the relationship between technology, social power, and human agency that arise from the smart city when we interpret it as Frankenstein's metropolis.

## Making a City Smart

The "smart city" is an urban planning and governance movement[1] that is predicated on widespread use of in-

---

[1] I refer to this as a "movement" to underscore the overarching aspects—discourses, ideologies, policies, and technologies—that tie together seemingly disparate or independent initiatives in different places around the world.

formation and communication technologies (ICT) to transform many aspects of the city. The technologies marshaled are often combinations of sensors, data analytics, algorithms, networked infrastructure, and operations centers, along with other ICTs as deemed necessary or useful. These technologies are built to work as a system: they are interlinked and interdependent, and together they allow for "smart" operations at a scale and type different from what any single ICT application could perform.

One such system, the Intelligent Operations Center build by IBM in Rio de Janeiro, incorporates the data collected from various sources (e.g., sensors, cameras, meters, GPS, social media) about various elements (e.g., flows of people, things, and resources; weather conditions; and/or security and emergency reports). The data are then fused in central databases, analyzed using sophisticated software, and fed into algorithms that react automatically or provide recommendations to human decision makers. The operations center of the smart city — think of a NASA-esque control room, replete with massive displays, computer terminals, and technical personnel — acts like a cybernetic brain that monitors, manages, and manipulates the urban body. The networked and responsive infrastructure allows for real-time commands, while data collection over time aids analysis and action.

These technologies do not necessarily manifest in such an apparent, imposing way as the operations center. They also operate behind the scenes, like the so-called "predictive policing" systems used in many cities across the United States, which collect and process troves of data to create models of where crime is likely to happen and lists of people who are likely to be offenders. The idea is that police will then know where, and on whom, to focus their attention. However, what's collected and how it's processed is often a trade secret protected by the companies who build and sell these systems. The police might not even know how exactly the predictive systems they use work. Indeed, some reports show that not even the companies are fully aware of how their technology works, especially once machines start learning and changing on their own. For the creators and users of these technologies, being smart does not necessarily require understanding all the operations and implications of that smartness.

These technologies and the people operating them are core components of one commonly lauded vision of the smart city as an urban "system of systems." This system attempts to render the city legible and observable, subject to surveillance and control. As the geographer Rob Kitchin explains:

> [This] notion of a "smart city" refers to the increasing extent to which urban places are composed of "everyware" (Greenfield 2006); that is, pervasive and ubiquitous computing and digitally instrumented devices built into the very fabric of urban environments (e.g., fixed and wireless telecom networks, digitally controlled utility services and transport infrastructure, sensor and camera networks, building management systems, and so on) that are used to monitor, manage, and regulate city flows and processes, often in real-time, and mobile computing (e.g., smart phones) used by many urban citizens to engage with and navigate the city which themselves produce data about their users (such as location and activity).[2]

Kitchin refers to this as the "real-time city" to emphasize the speed at which data is collected and analyzed, and the responsiveness of managerial action based on that data (such as by remotely tweaking infrastructural flows or deploying agencies like police). A recent volume of work by technologists, architects, and artists explores an alternative label, the "sentient city," which emphasizes the

---

[2] Rob Kitchin, "The Real-Time City? Big Data and Smart Urbanism," *GeoJournal* 79 (2014): 1-2.

sensory and agential qualities of this type of city.[3] That is to say, the sentient city has the capacity for sensing stimuli, "knowing" what's happening, and acting (whether automatically or by human command) based on that sensory feedback. Others go so far as to refer to the layers of ICT as a "digital skin"—a sensory organ "composed of connected, digitally enabled objects, network nodes, communication devices and posts for monitoring and analyzing data fed into servers"[4]—that covers the urban environment.

## Frankenstein's Metropolis[5]

This vision of the smart city as *alive*, teeming with the energy of coupled human-ICT systems, invokes the idea of a Frankenstein creation. The characters are more complicated in this version of the story. But even when reality offers up analogs that are far more complex, morality tales like *Frankenstein* point to things we should be attentive to—in this case questions of creation, responsibility, control, and agency.

With Frankenstein's metropolis, there is no single creator whose duty it is to create with care and ensure good outcomes, or who can shoulder the consequences of neglect and irresponsibility. There is instead a constellation of actors whose motivations, wants, and ideals come to-

---

[3] Mark Shepard, *Sentient City: Ubiquitous Computing, Architecture, and the Future of Urban Space* (Cambridge, MA: MIT Press, 2011).

[4] Chirag Rabari & Michael Storper, "The Digital Skin of Cities: Urban Theory and Research in the Age of the Sensored and Metered City, Ubiquitous Computing and Big Data," *Cambridge Journal of Regions, Economy and Society* 8 (2015): 27.

[5] Parts of this section and the next are derived from Jathan Sadowski & Frank Pasquale, "The Spectrum of Control: A Social Theory of the Smart City," *First Monday* 20, no. 7 (2015).

gether to produce the smart city.[6] There are the corporations like IBM and Cisco that seek to sell the products and services, as well as the ideologies of technocratic progress, needed for (re)making cities into smart creations. There are the city leaders who need ways to invigorate their cities, and exciting innovations offer quick fixes for social, political, and economic ailments. There are the citizens who help birth the smart city by integrating the newly available smart capabilities into their lives and by contributing data, time, and money to the processes of smartification.

And the creation is not atomistic, self-contained, or singular in nature. We cannot easily draw boundaries around it. We cannot observe the whole thing. We cannot watch it act, take into account its past actions, and predict its next actions. This creation is composed of massive, sprawling, networked systems of technologies and people. It was created by people, but like Frankenstein's creature, it escapes its creators. It grows, transforms, and creeps beyond the creators' intentions and reach. The systems that give rise to the smart city are often invisible in different ways: they are immaterial, like data streams and WiFi signals; they are tangible but part of hidden infrastructure, like sensors and wires; and they are the unnoticed things we interact with during daily life, like card readers. The nature of this creation does not have the familiarity of form represented in *Frankenstein*. This, in turn, makes the smart city difficult to see, evaluate, challenge, and change. It is able to exert insidious control—gradual, subtle, overlooked, normalized—rather than blunt coercion. The effects of these smart systems are often dispersed and unequal. For some, the systems channel and amplify their

---

[6] Of course, each city will be somewhat different in its origins and outcomes, but for the sake of a larger point I will generalize here.

capacities, while for others the systems monitor and constrain their lives.

In Mary Shelley's novel, the creation confronts his creator and exclaims, "You are my creator, but I am your master;—obey!"[7] For the vast majority of people, the smart technologies embedded into the urban fabric follow this Frankensteinian logic, not usually through force, but in the sense that these systems shape urban society and influence people's lives. Whether it is the checkpoint that decides if you will be able to get on the subway or the algorithm that decides you are at risk of committing a crime, there is little opportunity to challenge this technological power. This power is not justified by being right and reliable. As Rolien Hoyng remarks, the actual smart city is characterized by failure, malfunction, and breakdown.[8] This gives its power an element of unpredictability and inconsistency, which all means the smart city is even more difficult to avoid or resist.

What's more, these smart systems gain efficacy because, perhaps now more than ever before, the boundaries between body, city, and technology are blurred. There are not so much discrete entities—the person, the building, the device—as there are cyborgian connections of flesh, concrete, and information. These entanglements should be thought of in relational terms. While they may not manifest as visceral biological hybrids, like a *Terminator*-style cyborg, the lack of viscera does not mean the entanglements are inconsequential or immaterial. We are not separate and independent from the technologies we use, the

---

[7] Mary Shelley, *Frankenstein; or The Modern Prometheus*, ed. D. L. Macdonald & Kathleen Scherf, 3rd Ed. (Peterborough, Ontario: Broadview Press, (2012) [1818]), 176.

[8] Rolien Susanne Hoyng, "From Infrastructural Breakdown to Data Vandalism: Repoliticizing the Smart City?" *Television & New Media* 17, no. 5 (2016): 397-415.

environments we live in, and the systems we rely on. They are part of who we are. As the city changes and becomes smarter, it also changes the inhabitants. The city dweller is better understood as an urban cyborg: one who does not live *in* the city, but who lives as *part* of the city. As Matthew Gandy puts it,

> *The emphasis of the cyborg on the material interface between the body and the city is perhaps most strikingly manifested in the physical infrastructure that links the human body to vast technological networks. If we understand the cyborg to be a cybernetic creation, a hybrid of machine and organism, then urban infrastructures can be conceptualized as a series of interconnecting life-support systems.*[9]

As people become urban cyborgs, bodies merged with cities, our interfaces with the system grow more entangled. Moreover, as Donna Haraway writes in her seminal "Cyborg Manifesto": "No objects, spaces, or bodies are sacred in themselves; any component can be interfaced with any other if the proper standard, the proper code, can be constructed for processing signals in a common language."[10] Part of Haraway's project was to map the large-scale "transitions from the comfortable old hierarchical dominations to the scary new networks," which she calls "informatics of domination." This term refers to how these ICTs are deployed and the values they designed with. Like any technology, their uses and consequences are the result of human choices and social context. It is entirely possible that these smart systems could become "informatics of liberation." Alas, that does not appear to

---

[9] Matthew Gandy, "Cyborg Urbanization: Complexity and Monstrosity in the Contemporary City," *International Journal of Urban and Regional Research* 29, no. 1 (2005): 28.

[10] Donna Haraway, "A Cyborg Manifesto: Science, Technology, and Socialist-Feminism in the Late Twentieth Century," in Donna Haraway, *Simians, Cyborgs, and Women: The Reinvention of Nature* (New York, NY: Routledge, 1991), 163.

be our current trajectory — in which decisions about who designs, uses, and benefits from the smart city are relegated to rarefied places like the executive boardroom and mayoral office. The rise of Frankenstein's metropolis doubles down on the dominating power of technologies that are increasingly interoperable, pervasive, and invasive.

With all the optimistic promises and hopeful visions surrounding "smart cities," it can be easy to lose track of the politics that are coded into these interconnected technologies and initiatives. If we conceptualize these urban transformations as merely neutral enhancements that bring unalloyed goods of efficiency and security, then we miss out on the socio-political, even ontological, aspects of what it means to be assimilated deeper into the functioning of the smart city. We become urban cyborgs, necessary nodes on networks of data, components installed into the city. The relationship goes both ways: the city's totalizing environment becomes an integral part of life as a smart urbanite.

The cyborgian connections of smart systems can be liberating, as people gain new capabilities and conveniences, which raise the standard of living and contribute to a better society. Or they can be dominating, as people are monitored and modulated, which ensures they are productive members of a control society. These two types — liberation and domination — are not incoherent or paradoxical; they exist as parallel realities inhabited by different people at the same time. Understanding the smart city, let alone transforming it, requires that we map out these disparities of benefit and harm.

**The Bite Back**

It is tempting to create gigantic systems and then marvel at their awesome power. How could the proud creators of Frankenstein's metropolis resist using their creation's capacity to the fullest extent? Yet, the bite back of unintended consequences — and the perversity of intended outcomes — does not always follow a karmic cycle wherein people get what they deserve. In the novel, Victor Frankenstein is eventually ruined by his creation, his life thrown into shambles. But in reality, the creators gain more power from the systems they build and wield, further extending and entrenching their positions in and over society.

When smart ICT is touted as a way to understand and manage society, it is set to rationalize unjust patterns of discipline and control. For example, as a recent study shows, Chicago's predictive policing system was not the paragon of smartness that its proponents claimed it was. Instead, it reproduced existing biases, created new forms of profiling, and targeted people unnecessarily.[11] The "science of society" promised by the creators of smart cities morphs into a subjugation of certain parts of society. The software and algorithms behind such judgments become less objective arbiters of opportunity and punishment, and more ways of laundering subjective, biased decisions into initiatives touted as neutral, apolitical, and pragmatic.

Upon reflection, perhaps Frankenstein's metropolis is more like a Golem than a runaway monster. That is to say, it is out of control for those (i.e., most of us) who do not program its processes, access its inner workings, and direct its operational logics. The irony is that "Golem," in Modern Hebrew, translates to "dumb." So perhaps the city that we create may not be so "smart" after all.

---

[11] Jessica Saunders, Priscilla Hunt, and John S. Hollywood, "Predictions Put into Practice: A Quasi-Experimental Evaluation of Chicago's Predictive Policing Pilot," *Journal of Experimental Criminology* 12, no. 3 (2016): 347-371.

# 10

## MONSTER MYTHOS: *FRANKENSTEIN* AS NETWORK TEXT

## Ed Finn

Few stories of the modern era have thrived so effectively as Mary Shelley's *Frankenstein; or, The Modern Prometheus*. First published in 1818, the novel immediately became a popular sensation, with numerous stage adaptations and endless gossip about its authorship and its moral character. After all, Shelley had written a tale that was bright with the latest scientific discovery and dark with Gothic sentiment. Victor Frankenstein and his creature are doomed twins, wrestling with the consequences of their actions in high Romantic style. Shelley's work was shocking precisely because it refused to answer the questions it raised: as the *Quarterly Review* put it in 1817, the novel "inculcates no lesson of conduct, manners or morality."[1] At the same time, the story was too interesting to simply

---

[1] Quoted in Susan Tyler Hitchcock, *Frankenstein: A Cultural History* (New York: W. W. Norton & Co., 2007). My account of the novel's early reception is indebted to her excellent cultural history and to D.L. Macdonald and Kathleen Scherf's edition of *Frankenstein; or, The Modern Prometheus*, 3rd ed. (Peterborough, Ontario: Broadview Press, 2012 [1818]).

dismiss. It brought together sophisticated philosophical thought, the latest scientific theories on electricity and the nature of life, and nuanced layers of literary reference. When Walter Scott reviewed the novel in the esteemed *Blackwood's Edinburgh Magazine*, he praised its "uncommon powers of poetic imagination" and celebrated it as a higher form of literature, foreshadowing its science fiction genre capacity "to open new trains and channels of thought."[2]

While Scott gave *Frankenstein* and its then-anonymous author his warm endorsement, other reviewers argued that its lack of a clear moral argument made it "an uncouth story," "almost ... impious" and certainly, if the rumors were true that the author was female, an inexcusable failure to maintain "the gentleness of her sex."[3] As Shelley soon realized, the bad reviews were almost as helpful as the good ones in popularizing the novel, which quickly sold out its first edition. What followed was the first indication that Shelley had struck upon an idea that did not merely draw together a diverse set of intellectual energies but transformed them into a kind of viral narrative. Within five years of the novel's publication it had been adapted for the stage five times, and only a year later made its first appearance as a broader cultural referent in a parliamentary debate about emancipating British slaves.[4]

Something unique happened here: from the beginning, the novel was both highly allusive and productively am-

---

[2] Walter Scott, "Remarks on *Frankenstein, or the Modern Prometheus*; a Novel"; see also Hitchcock, 75.

[3] Reviews from *The Quarterly Review, Bell's Court and Fashionable Magazine,* and *British Critic* quoted in Hitchcock, 74-76.

[4] Hitchcock, 88-89. See also Elizabeth Young's *Black Frankenstein: The Making of an American Metaphor* (New York, NY: NYU Press, 2008).

biguous, offering many pathways of interpretation to its readers and many opportunities for adaptation. Despite having the odds stacked against her because of her gender, her social ignominy for her unconventional relationship with Percy Bysshe Shelley, and the outlandish subject of her novel, Shelley published a story that transformed its cultural sphere in a few short years.

In this essay I argue that *Frankenstein* is not merely an early example of science fiction, but a "network text." I use this term to draw together several different discourses about the role of text as information. In the social sciences, network text analysis is a term used to describe how text can reveal networks of relationships between people, institutions, and ideas through measures such as mutual information and co-occurrence.[5] In the humanities, the term has roots to scholarship about hypertext literature and the notion of digital literature that encodes certain forms of literal connection, as well as allusion and metaphor.[6] In the study of an increasingly complex, boundary-defying digital culture, the presence or idea of a network, with all its informational, sociological, and literary historical implications, has become almost ubiquitous, and in many ways the subject of novelists itself, from Thomas Pynchon's *Bleeding Edge* to Dave Eggers's *The Circle*.

While Shelley's novel did not rely on HTML or encoded linking structures, it nevertheless referenced and transformed a rich web of ideas, conversations, and intellectual

---

[5] Roel Popping, *Computer-assisted Text Analysis* (Thousand Oaks, CA: SAGE Publications, 2000).

[6] See George P. Landow, *Hypertext 3.0: Critical Theory and New Media in an Era of Globalization*, 3rd Ed. (Baltimore, MD: Johns Hopkins University Press, 2006); N. Katherine Hayles, "The Transformation of Narrative and the Materiality of Hypertext," *Narrative* 9, no. 1 (2001): 21-39; and David Ciccoricco, *Reading Network Fiction* (Tuscaloosa, AL: University of Alabama Press, 2007).

ley lines. In many ways, Shelley was grappling with the cutting edges of the Enlightenment, testing the limits of socially acceptable love, faith, and ethics in the aftermath of the French and American revolutions, the rise of the industrial age and, most importantly, the emergence of a modern cultural economy. These tangled webs are not just the raw materials for Shelley's work, but the sinews and sutures of the monster itself.

Through allusions, epigraphs, a dedication, in-text bibliographies, and various other means, *Frankenstein* traces out this sophisticated network architecture, weaving together the era's philosophical, scientific, moral, and religious debates in a work that was both intellectually provocative — almost shocking — for its time, and written for a popular audience in the still-somewhat-disreputable genre of the novel. By the time of the 1831 edition it seems fair to say that the story had already achieved mythic status, in the sense that almost everyone in English society (if not elsewhere, as translations, adaptations, and pirated works proliferated) knew the story of Victor Frankenstein and his creature, whether or not they ever read Shelley's novel.

This highly successful diffusion proceeded in parallel with the novel's rapid canonization. The 1831 edition of the novel introduced substantial changes (the subject of much literary scholarship), which Shelley introduced in part to meet the expectations of the new literary commodity brand it would be joining. *Frankenstein* was reissued in the Bentley's Standard Novels series, which were mostly published as "Revised, Corrected, and Illustrated, with a New Introduction by the Author," and Shelley took the opportunity to significantly edit her work and author a now-famous preface articulating the book's own creation myth.[7] Like the Book-of-the-Month Club described by Janice Radway a century later, the publisher Richard Bentley had created a standardized literary experience that included recent celebrated authors like Jane Austen and Edward Bulwer-Lytton.[8]

This commercialization of *Frankenstein* both elevated it, by giving it the same cultural middlebrow value as the other identically bound and produced novels in the series, and constrained it, through Bentley's long stranglehold on the novel's copyright, which kept *Frankenstein* out of print for much of the latter half of the 19th century.[9] To put it another way, Bentley broadly disseminated the idea of *Frankenstein* while restricting direct access to the text through upmarket print editions targeted at a well-heeled bourgeois audience. As William St. Clair puts it:

> *Refused a life in the reasonably stable culture of print and reading,* Frankenstein *survived in a free-floating popular oral and visual culture, with only the central episode of the scientist making the Creature holding it tenuously to its original.*[10]

Little wonder that the Frankenstein narrative spread through other means, permeating popular consciousness through adaptations and allusions of all kinds, and rapidly becoming its own cliché about the hazards of change and the hubris of science.

---

[7] William St. Clair, "The Impact of *Frankenstein*," in Betty T. Bennett & Stuart Curran, eds., *Mary Shelley in Her Times* (Baltimore, MD: Johns Hopkins University Press, 2000), 44.

[8] Janice A. Radway, *A Feeling for Books: The Book-of-the-Month Club, Literary Taste, and Middle-Class Desire* (Chapel Hill, NC: University of North Carolina Press, 1997).

[9] St. Clair, 46-49.

[10] St. Clair, 54.

One of the most remarkable of these early versions of the network text was an 1849 stage play titled *Franken- stein, or the Model Man*, in which the creature was a "mechanical man with skill supreme." This robotic monster cross-pollinated the myth of the secret of life with the threat of mechanization, offering a compelling example of how myths recombine and reproduce in popular culture. But this monster goes a step further, in that it is constructed not merely from the expected components of Shelley's novel, the burgeoning industrial revolution, previous iterations of the Frankenstein story, and English stage traditions.

As Susan Tyler Hitchcock reports in her cultural history of *Frankenstein*, the model man is "brought to life with a potion, half alchemical 'Elixir Vitae' and half Victorian-era wonder drug," assembled out of the miracle cures and snake oil advertisements of the time.[11] In other words, this was a creature of advertising—of Slolberg's lozenge, Grimstone's Aromatic Regenerator, Parr's life pill, and other branded ingredients teeming with consumer vitality. This is a kind of alchemy through replication. The many copies of these advertisements that the play's audience would have seen in newspapers and serials were now copied once again, but also transmuted into living flesh and willful agency.

These advertisements formed a rich network of paratext surrounding, infiltrating, and sometimes contradicting the Victorian public sphere, reproducing brands, images, and consumer ideals across a range of different publications, social strata, and geographical areas. To make them into the ingredients for creating life acknowledged the fundamentally mediated nature of this mechanical creature a century before Marshall McLuhan or cybernetic theory. *Frankenstein* had already become a con-

---

[11] Hitchcock, 102.

duit for ideas, a living metaphor that could be adapted and recast to new political, cultural, and economic situations, animated and reanimated by advertisements and commercial reinvention.

This version of Frankenstein's creation was not just assembled out of advertisements, as a performance on stage, but out of commodities themselves: an embodiment of the early networked capitalism of the Victorian era. As Andrew Burkett has argued, Shelley's linkage between mediation and monstrosity vitalized and personified the more abstract philosophical debate about the relationship between information and materiality.[12] *Frankenstein* was born of the networks of Romantic science and imagination surrounding Mary Shelley since her childhood, but it also gave birth to a creature who *is* the network, assembled from pieces of Locke and Rousseau, Milton and Plutarch, Christianity and Greek myth, philosophy and poetry, of book contracts and advertisements, of stage plays and political cartoons.

### The Creature Flourishes

As the myth continued to grow and diversify, *Frankenstein* came back into print in 1880.[13] By the dawn of the film era, Victor Frankenstein and his creature were firmly established in popular culture as familiar signifiers that could be attached to almost any line of argument. Their depiction in an early Thomas Edison film once again depended on the close manipulation of medium in order to give new life to the creature. The film's special effects included a shot of a burning effigy played backwards, al-

---

[12] Andrew Burkett, "Mediating Monstrosity: Media, Information, and Mary Shelley's *Frankenstein,*" *Studies in Romanticism* 51, no. 4 (2012): 579-605.

[13] St. Clair, 49.

lowing Frankenstein's monster to assemble itself from a smoking cauldron. Once again the monster was assembling itself out of media, emerging not just from the fires of Victor Frankenstein's presumption but from the fragile celluloid used to produce most films of the era. The highly combustible nitrate stock, which led to a number of fatal fires, is spliced backwards so that smoke drifts *down* into the massive cauldron from which the effigy seems to emerge. The temporal reversal offers a neat commentary on the Edison film's role in projecting a kind of science fictional future by descending into the past. The smoke from the cauldron appears to move backwards, as if being pulled into the simmering, ashy cauldron of cultural imagination, where mediated myths continually cycle through death and rebirth. Perhaps, as Frederick C. Wiebel speculates, the effigy was so flammable because it was literally wrapped in nitrate film, once again creating a network monster out of media itself.[14]

### IMDb Titles with Keyword "Frankenstein"

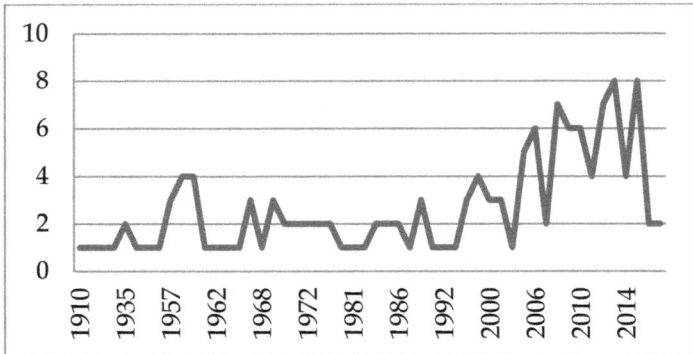

The Edison film was just the first in a long stream of *Frankenstein* adaptations, reinventions, and references in

---

[14] Frederick C. Wiebel, Jr., *Edison's Frankenstein* (Albany, GA: BearManor Media, 2010).

film history. Arguably the most famous Frankenstein film of all, Boris Karloff's unforgettable presentation of the creature, is so deeply covered by contemporary cultural criticism that I'll merely note that it functioned as a second ignition for the film trajectory of *Frankenstein*, creating an indelible visual image to flesh out the monster in popular culture. Since James Whale's production in 1931, there have been at least 150 *Frankenstein*-related productions, according to the Internet Movie Database. And that number continues to grow.

It is striking to see not just the gradual rise in *Frankenstein* films (which is likely explained in part by a general increase in the number of film productions each year overall), but the way in which the increase cycles between particularly active years, like 2008 or 2015. These cycles of interest and fame are familiar to me from my work on authors like Thomas Pynchon and the Pynchon industry. Because cultural conversations are themselves network texts, I believe these cycles of attention offer a glimpse into how certain memes and ideas can travel through cascades of information—not just one new *Frankenstein* project but a handful, all responding somehow to one another and the same broader matrix of cultural discourse. A good myth is a cultural safe haven, a proven vehicle for shuttling new ideas into the cultural consciousness or hedging your bets at the box office. We can think of this as the buffering quality of metaphor, allowing us to approach deeply alarming subjects through the protective membrane of fiction, allegory, and analogy. This certainly aligns with Frankenstein's persistent deployment to contend with all sorts of *others*, from the first parliamentary debate about the hazards of emancipating West Indian slaves to Michael Jackson's "Thriller" a century and a half later. It is also a vital function of fiction, particularly science fiction, allowing authors and readers to consider "unthinkable" subjects.

More recent treatments of *Frankenstein* continue to explore this vital othering role for the myth, from Elizabeth Young's remarkable history of race and the other in *Black Frankenstein* to the queering of the creature in Shelley Jackson's masterful electronic work *Patchwork Girl*. As we approach the novel's bicentennial in 2018, it's clear that Shelley's story remains vital as both a unifying story, a blockbuster myth if you will, and as a pathway for subaltern revisions and adaptations that use its very mainstream iconicity as a way to frame minority questions in ways the majority cannot easily ignore.

This is the context in which *Frankenstein* influences another modern network creature, the luckless protagonist of Junot Díaz's masterful *The Brief Wondrous Life of Oscar Wao*. Oscar is a Dominican-American boy trapped between the science fiction and fantasy culture of mainstream American nerd-dom and the searing colonial history of the Trujillo regime: the network he inhabits owes more to J. R. R. Tolkein and Marvel Comics than to galvanism, but the mediated other who speaks from the page is deeply familiar.[15] As Díaz explained, Oscar is "a Frankenstein monster in the sense that he's not a monster or beast, but he's the subject or person that someone else is trying to piece together."[16] Díaz's novel nudges his readers into exactly this kind of reconstruction through a rich web of allusions to popular culture, 4-color characters from comics and cult films, as well through an extended history lesson on the Dominican Republic, and the persistent use of unglossed Spanish. The reader must assemble her own creature, a narrative and protagonist assembled

[15] Ed Finn, "Revenge of the Nerd: Junot Díaz and the Networks of American Literary Imagination," *Digital Humanities Quarterly* 7, no. 1 (2013).

[16] Evelyn Ch-ien, "The Exploding Planet of Junot Díaz," *Granta* (27 April 2008).

out of the fragments of multiple cultures colliding in the Caribbean and the United States.

## Remaking the Monster

The theme of participatory making, of the audience putting in significant work to make the creature whole, extends back to the original framing of the novel. Shelley's epigraph from John Milton's *Paradise Lost* ("Did I request thee, maker, from thy clay to mold me man?") is richly productive in its own right, but one central theme is the responsibility of creation. In Milton's original epic, Adam imagines the unfolding consequences of his own procreation, the generations of humanity blaming him for his original sin. But by placing the quotation out of context, Shelley puts the question on a neutral footing, pointing out the deeper question of when, and why, such acts of creation are justified.

As *Frankenstein* continues to reinvent and revitalize cultural imaginations in the 21st century, the question of collaborative making will become increasingly important to new iterations of the myth. Cultural production is much less centralized than it was in the era of 19th century theatrical adaptations or even 20th century Hollywood films. The increasingly tenuous distinctions between consumption and production in the digital era only amplify the invitation implicit in the *Frankenstein* myth to take this network monster and rebuild him (or her) for yourself. Social media platforms, fan culture, the so-called sharing economy, and the growth of ad-hoc labor markets all push today's readers to also be writers and our audience-goers to be artists. We all perform *Frankenstein* now, and therefore we perform its central ethical questions. The cosplayer designing a monstrous new costume and the child performing the iconic creature's lurch in an imagination game at home both confront the status of the other in the ways that they enact the myth: friendly creature or scary monster; articulate being or mute creation; cultural participant or social outcast.

Perhaps it is inevitable that the monster takes on new material forms to echo its historical constitution through human corpses, Victorian advertisements, and Hollywood special effects. It is a network myth, and today the network defies distinctions between virtual and real as ubiquitous computation, cheap sensors, and algorithmic platforms from Uber to Google Now transform the lived experience of reality through algorithmic interventions. *Frankenstein* began as a kind of network effect, distilling undercurrents of political, scientific, aesthetic, and moral revolutionary sentiments: it was a young book for a new age, throwing off conventions to present a narrative that was both highly topical to its era (and Mary Shelley's tumultuous personal experiences) and written for a much broader stage of literary history, encompassing Plutarch, Milton, and the passage of empires. As the story has evolved from effect to a network entity in its own right, it has recreated the conditions of its telling across many media and narrative forms, as if the assemblage of the mediated monster follows the logic of biological development: ontology recapitulates phylogeny.

Such a diversity of materials is a survival tactic. Myths are hard to kill precisely because of their replicating, chimeric nature—the different systems of limbs and limbic impulses that we thread together into stories to frighten and delight. They are truly supernatural stories, untethered from the obligations of mortal narratives: setting, plot, or climax. They are ideas endowed with agency and characterization. The monster haunts us because he is both a child and a parent of ideas, stitched together in the dark corners of our collective imaginations and animated by the spark, the vital energy, of deep-seated hopes and fears. The sutures and rough stitching that we imagine holding Victor's creature together are just reminders for the real threading that links together the best and worst of the Enlightenment's long history. The monster continues to animate science and scientists, along with many others working at the intersection of creativity and responsibility. We need the networked creature to connect intellectual responsibility with the pride and peril of parenthood, bridging the poetic and the scientific imagination. Frankenstein is a network creature, an exquisite corpse, a body of media artifacts, ideas, consumer products, and performed deaths assembled by many hands over two centuries of fraught history.

# AFTERWORD: VOLCANOES, MONSTERS, AND POLITICAL ECOLOGY

The story of Victor Frankenstein and his creature was born during an unnatural winter in June. The Year Without a Summer was part of a larger climatic event, "the Little Ice Age" that began around 1300 CE and continued into the 19th century. Thanks to the eruption of Mount Tambora, temperatures dipped even further to wreak havoc on human societies.[1] The famines, riots, and diseases that came in the wake of this series of natural events may seem all too familiar as we face down hurricanes and flooding on one of the U.S. coasts, and drought and wildfires on the other. This geologic age has been dubbed the Anthropocene because the natural disasters we face are shaped by the collective actions of humans. If these new global threats are the creature we have created together, then we are tasked with the same rigorous ethical thinking we've asked of Victor Frankenstein in these chapters.

In our introduction, we quoted Bruno Latour, who stressed the urgency of loving our monsters. In the same

---

[1] Brian Fagan, *The Little Ice Age: How Climate Made History 1300-1850* (New York, NY: Basic Books, 2001).

essay, Latour challenges us to look for these monsters in our relationship to nature.

> *Let Dr. Frankenstein's sin serve as a parable for political ecology. At a time when science, technology, and demography make clear that we can never separate ourselves from the nonhuman world — that we, our technologies, and nature can no more be disentangled than we can remember the distinction between Dr. Frankenstein and his monster — this is the moment chosen by millions of well-meaning souls to flagellate themselves for their earlier aspiration to dominion, to repent for their past hubris, to look for ways of diminishing the numbers of their fellow humans, and to swear to make their footprints invisible?[2]*

Latour's idea of political ecology here is a break from more traditional ideas of political ecology, which call for a politics that cares for nature by leaving it alone. Instead, he suggests, we must embrace the idea that science, technology, nature, and culture are "so confused and mixed up as to be impossible to untangle." When we embrace this idea, our politics does not protect the pure, natural world from us, but considers this mangle of human and nonhuman, natural and unnatural when making decisions about how to live and move through the world. By stretching the Frankenstein metaphor to include a creation that extends so far beyond a single, corporeal creature, Latour pushes us to consider responsibility and creativity not as individual traits, but as part of the fabric of our culture — to consider our communal responsibility for our shared creative endeavors. He continues:

> *The goal of political ecology must not be to stop innovating, inventing, creating, and intervening. The real goal must be to have the same type of patience and commitment to our*

---

[2] Bruno Latour, "Love Your Monsters: Why We Must Care for Our Technologies As We Do Our Children," *The Breakthrough Journal* (Winter 2012).

*creations as God the Creator, Himself. And the comparison is not blasphemous: we have taken the whole of Creation on our shoulders and have become coextensive with the Earth.*[3]

Political ecology may be an important way for western culture to embrace coexistence and respect for nature precisely because part of our heritage is built on carelessness and ambition. *Frankenstein* is one of our enduring myths because we know Victor Frankenstein. We know a thousand Victor Frankensteins. Other cultures operate from their own sets of narratives, beliefs, and practices. Many indigenous cultures have developed a close, interdependent relationship with the natural world. What Latour calls political ecology is already an established part of many other worldviews. As we invest deeply in Frankenstein's story as a metaphor for our relationships with science and technology or with each other, we should be mindful that some cultures have no need for a mythology that teaches them about the peril they face when they fail to think before they leap or to care for what they create. Not all cultures need a narrative like Shelley's to help them make sense of their role in the world, but it seems we do.

We are in danger, though, of failing to hear the story Mary Shelley is telling us. It would be a mistake to abandon our creations or fail to care for them. But it would also be a mistake to abandon innovation. The creature was not inexorably destined for murder and mayhem. He was abandoned to it. If loving our monsters means opening ourselves to political ecology, then the onset of the Anthropocene means we have a lot more work to do.

— *Megan Halpern*

---

[3] Latour.

# ABOUT THE AUTHORS

**Joey Eschrich** is the editor and program manager for the Center for Science and the Imagination at Arizona State University, coeditor of *Everything Change: An Anthology of Climate Fiction* (2016), and managing editor of *Frankenstein: Annotated for Scientists, Engineers, and Creators of All Kinds* (2017).

**Kevin M. Esvelt** is an assistant professor and head of the Sculpting Evolution Group at the Massachusetts Institute of Technology Media Lab.

**Ed Finn** is an assistant professor and the founding director of the Center for Science and the Imagination at Arizona State University. He is the author of *What Algorithms Want: Imagination in the Age of Computing* (2017), and coeditor of *Frankenstein: Annotated for Scientists, Engineers, and Creators of All Kinds* (2017).

**Charlotte Gordon** is the Distinguished Professor of Humanities at Endicott College. Her latest book, *Romantic Outlaws: The Extraordinary Lives of Mary Wollstonecraft and Mary Shelley* (2015), won the National Book Critics Circle award.

**David Guston** is Foundation Professor and founding director of the School for the Future of Innovation in Society at Arizona State University, where he is also codirector of the Consortium for Science, Policy & Outcomes. He is the founding editor of the *Journal of Responsible Innovation* and coeditor of *Frankenstein: Annotated for Scientists, Engineers, and Creators of All Kinds* (2017).

**Megan Halpern** is an assistant professor at Michigan State University in the History, Philosophy, and Sociology of Science at Lyman Briggs College and the Residential College for Arts and Humanities.

**Jathan Sadowski** is a postdoctoral research fellow in smart cities at the University of Sydney. He has written for *The Guardian*, *The Nation*, and *Slate*, among other outlets.

**Alyssa Sims** is a program associate with the International Security program at New America.

**Bina Venkataraman** is a fellow at New America and teaches at the Massachusetts Institute of Technology. She is the director of global policy initiatives at the Broad Institute of Harvard and MIT.

**Sara Imari Walker** is a theoretical physicist, astrobiologist, and assistant professor in the School of Earth and Space Exploration at Arizona State University.

# ACKNOWLEDGEMENTS

The editors would like to thank all of the contributors to this volume for their fresh perspectives on *Frankenstein*, one of the world's most thought-about and written-about works of literature. Everyone was incredibly generous with their time and energy and weathered our edits and questions with grace. Thank you all for making this such a fun and intellectually stimulating process.

We'd also like to thank G. Pascal Zachary, series editor for *The Rightful Place of Science*, for his guidance and good counsel, enthusiasm for this collection, and vision in creating a unique book series that has created a new space for lucid, concise, challenging public conversations about science, technology, and society.

Several of the chapters in this book are adapted from articles that first appeared in *Slate Magazine*. We express sincere appreciation to *Slate* for agreeing to let us reprint the pieces by Bina Venkataraman, Charlotte Gordon, Kevin Esvelt, and Alyssa Sims.

Thanks are due to the multitalented Jason Lloyd, who was instrumental in shepherding this book into the world and in improving our work with incisive questions and assiduous wordsmithing. Thanks also for the Frankenstein's creature glyph by Arianna Sbaffi from the Noun Project. And we are grateful for funding from the National Science Foundation (award #0937591), which supported Megan Halpern during her postdoctoral research at the

Center for Nanotechnology in Society at Arizona State University.

We'd like to specially thank two of our contributors, David H. Guston and Ed Finn, for their able leadership of the Frankenstein Bicentennial Project at Arizona State University. Without their vision and support—and without their forethought years before the bicentennial—none of us would be working on and learning from *Frankenstein* today.

www.ingramcontent.com/pod-product-compliance
Lightning Source LLC
Chambersburg PA
CBHW060620210326
41520CB00010B/1415